Selected Titles in This Series

747 **Armand Borel, Robert Friedman, and John W. Morgan,** Almost commuting elements in compact Lie groups, 2002

746 **Peter Niemann,** Some generalized Kac-Moody algebras with known root multiplicities, 2002

745 **Mikhail A. Lifshits and Werner Linde,** Approximation and entropy numbers of Volterra operators with application to Brownian motion, 2002

744 **Roger Chalkley,** Basic global relative invariants for homogeneous linear differential equations, 2002

743 **Heng Sun,** Spectral decomposition of a covering of $GL(r)$: the Borel case, 2002

742 **J. E. Gilbert, Y. S. Han, J. A. Hogan, J. D. Lakey, D. Weiland, and G. Weiss,** Smooth molecular functions and singular integral operators, 2002

741 **Francisco Santos,** Triangulations of oriented matroids, 2002

740 **Rick Durrett,** Mutual invadability implies coexistence in spatial models, 2002

739 **Georgios K. Alexopoulos,** Sub-Laplacians with drift on Lie groups of polynomial volume growth, 2002

738 **Yasuro Gon,** Generalized Whittaker functions on $SU(2,2)$ with respect to the Siegel parabolic subgroup, 2002

737 **Arjen Doelman, Robert A. Gardner, and Tasso J. Kaper,** A stability index analysis of 1-D patterns of the Gray-Scott model, 2002

736 **Wojciech Chachólski and Jérôme Scherer,** Homotopy theory of diagrams, 2002

735 **Martina Brück, Xi Du, Joonsang Park, and Chuu-Lian Terng,** The submanifold geometries associated to Grassmannian systems, 2002

734 **Michel Van den Bergh,** Blowing up of non-commutative smooth surfaces, 2001

733 **Milé Krajčevski,** Tilings of the plane, hyperbolic groups and small cancellation conditions, 2001

732 **Jan O. Kleppe, Juan C. Migliore, Rosa Miró-Roig, Uwe Nagel, and Chris Peterson,** Gorenstein liaison, complete intersection liaison invariants and unobstructedness, 2001

731 **Jesús Bastero, Mario Milman, and Francisco J. Ruiz,** On the connection between weighted norm inequalities, commutators and real interpolation, 2001

730 **Suhyoung Choi,** The decomposition and classification of radiant affine 3-manifolds, 2001

729 **Michael Grosser, Eva Farkas, Michael Kunzinger, and Roland Steinbauer,** On the foundations of nonlinear generalized functions I and II, 2001

728 **Laura Smithies,** Equivariant analytic localization of group representations, 2001

727 **Anthony D. Blaom,** A geometric setting for Hamiltonian perturbation theory, 2001

726 **Victor L. Shapiro,** Singular quasilinearity and higher eigenvalues, 2001

725 **Jean-Pierre Rosay and Edgar Lee Stout,** Strong boundary values, analytic functionals, and nonlinear Paley-Wiener theory, 2001

724 **Lisa Carbone,** Non-uniform lattices on uniform trees, 2001

723 **Deborah M. King and John B. Strantzen,** Maximum entropy of cycles of even period, 2001

722 **Hernán Cendra, Jerrold E. Marsden, and Tudor S. Ratiu,** Lagrangian reduction by stages, 2001

721 **Ingrid C. Bauer,** Surfaces with $K^2 = 7$ and $p_g = 4$, 2001

720 **Palle E. T. Jorgensen,** Ruelle operators: Functions which are harmonic with respect to a transfer operator, 2001

719 **Steve Hofmann and John L. Lewis,** The Dirichlet problem for parabolic operators with singular drift terms, 2001

(*Continued in the back of this publication*)

Almost Commuting Elements in Compact Lie Groups

MEMOIRS
of the
American Mathematical Society

Number 747

Almost Commuting Elements in Compact Lie Groups

Armand Borel
Robert Friedman
John W. Morgan

May 2002 • Volume 157 • Number 747 (third of 5 numbers) • ISSN 0065-9266

American Mathematical Society
Providence, Rhode Island

2000 *Mathematics Subject Classification.*
Primary 22C05, 17B20; Secondary 57R20, 17B40.

Library of Congress Cataloging-in-Publication Data

Borel, Armand.
Almost commuting elements in compact Lie groups / Armand Borel, Robert Friedman, John W. Morgan.
p. cm. — (Memoirs of the American Mathematical Society, ISSN 0065-9266 ; no. 747)
"Volume 157, number 747 (third of 5 numbers)."
Includes bibliographical references.
ISBN 0-8218-2792-8 (alk. paper)
1. Lie groups. 2. Compact groups. 3. Root systems (Algebra) I. Friedman, Robert, 1950– II. Morgan, John W., 1946– III. Title. IV. Series.

QA3 .A57 no. 747
[QA387]
510 s—dc21
[512′.55] 2002018237

Memoirs of the American Mathematical Society

This journal is devoted entirely to research in pure and applied mathematics.

Subscription information. The 2002 subscription begins with volume 155 and consists of six mailings, each containing one or more numbers. Subscription prices for 2002 are $524 list, $419 institutional member. A late charge of 10% of the subscription price will be imposed on orders received from nonmembers after January 1 of the subscription year. Subscribers outside the United States and India must pay a postage surcharge of $31; subscribers in India must pay a postage surcharge of $43. Expedited delivery to destinations in North America $35; elsewhere $130. Each number may be ordered separately; *please specify number* when ordering an individual number. For prices and titles of recently released numbers, see the New Publications sections of the *Notices of the American Mathematical Society.*

Back number information. For back issues see the *AMS Catalog of Publications.*

Subscriptions and orders should be addressed to the American Mathematical Society, P. O. Box 845904, Boston, MA 02284-5904. *All orders must be accompanied by payment.* Other correspondence should be addressed to Box 6248, Providence, RI 02940-6248.

Memoirs of the American Mathematical Society is published bimonthly (each volume consisting usually of more than one number) by the American Mathematical Society at 201 Charles Street, Providence, RI 02904-2294. Periodicals postage paid at Providence, RI. Postmaster: Send address changes to Memoirs, American Mathematical Society, P. O. Box 6248, Providence, RI 02940-6248.

This publication is indexed in *Science Citation Index®*, *SciSearch®*, *Research Alert®*, *CompuMath Citation Index®*, *Current Contents®/Physical, Chemical & Earth Sciences.*
Printed in the United States of America.

∞ The paper used in this book is acid-free and falls within the guidelines established to ensure permanence and durability.
Visit the AMS home page at URL: `http://www.ams.org/`

10 9 8 7 6 5 4 3 2 1 07 06 05 04 03 02

Contents

Abstract

We describe the components of the moduli space of conjugacy classes of commuting pairs and triples of elements in a compact Lie group. This description is in terms of the extended Dynkin diagram of the simply connected cover, together with the coroot integers and the action of the fundamental group. In the case of three commuting elements, we compute Chern-Simons invariants associated to the corresponding flat bundles over the three-torus, and verify a conjecture of Witten which reveals a surprising symmetry involving the Chern-Simons invariants and the dimensions of the components of the moduli space.

Received by the editor October 4, 1999, and in revised form August 28, 2000.
2000 *Mathematics Subject Classification.* Primary 22C05, 17B20; Secondary 57R20, 17B40.
Key words and phrases. Lie group, root system, Chern-Simons.
The second author was supported in part by NSF Grant #DMS-96-22681.
The third author was supported in part by NSF Grant #DMS-97-04507.

CHAPTER 1

Introduction

Let K be a compact, connected and semisimple Lie group. This paper describes the moduli space of isomorphism classes of flat connections on principal K-bundles over the two-torus and the three-torus. There are two motivations for this study. The first is the relation between flat K-bundles over the two-torus and holomorphic principal bundles over an elliptic curve with structure group the complexification of K. The second is to give a proof of a conjecture of Witten concerning the moduli space of flat K-bundles over the three-torus, used in [**20**].

Of course, a flat bundle is completely determined by its holonomy representation, so that the problem of classifying flat bundles over two- and three-tori becomes the question of classifying ordered pairs and triples of commuting elements in K, up to simultaneous conjugation. If two elements commute in K, then any lifts of them to the universal cover of K commute up to an element of the center. Thus we shall work in the simply connected cover G of K and study pairs and triples of elements in G which commute up to the center, hence the name "almost commuting." This is the form in which we attack the question. Our point of view is that the extended Dynkin diagram of G, the action of $\pi_1(K)$ on this diagram, and the coroot integers associated to this diagram completely determine the answer in a manner which we shall describe below.

Notation used throughout this paper

• G is a compact, connected and simply connected Lie group, and in particular G is semi-simple, T is a maximal torus of G, and $\mathfrak{t} = \mathrm{Lie}(T)$.

• $\mathcal{C}G$ denotes the center of G. If G is simple, then $\mathcal{C}G$ is cyclic except for $G = Spin(4n)$, $n \geq 2$. An element of $\mathcal{C}Spin(4n)$ whose image in $SO(4n)$ is nontrivial will be called *exotic*.

• For any subset $X \subseteq G$, $Z_G(X)$ denotes the centralizer of X in G. We will denote $Z_G(X)$ by $Z(X)$ when G is clear from the context.

• If S is a torus in G, not necessarily maximal, we let $W(S, G)$ be the finite group $N_G(S)/Z_G(S)$ and we call it the *Weyl group of S in G*. If $\mathfrak{s} = \mathrm{Lie}(S)$, then we set $W(\mathfrak{s}, G) = W(S, G)$.

• Given $x, y \in G$, we define the commutator $[x, y] = xyx^{-1}y^{-1}$ and denote conjugation by x as $i_x(y) = {}^x y = xyx^{-1}$.

One convention concerning subtori used throughout the paper is the following. Let $\mathfrak{t}$ be a vector space with a positive definite inner product $\langle \cdot, \cdot \rangle$, let $\Lambda \subseteq \mathfrak{t}$ be a lattice, such that $\langle v, w \rangle \in \mathbf{Z}$ for all $v, w \in \Lambda$, and let $T = \mathfrak{t}/\Lambda$ be the associated torus. For any subtorus $S \subseteq T$ let $\mathfrak{s} \subseteq \mathfrak{t}$ be its tangent space. Let $\mathfrak{s}^{\perp}$ be the perpendicular subspace to $\mathfrak{s}$. Then $\mathfrak{s}^{\perp}$ is the tangent space of another subtorus $S' \subseteq T$ and $F_S = S \cap S'$ is a finite group. We denote by $\overline{S}$ the quotient torus

$S/F_S = T/S'$. Clearly $S = \mathfrak{s}/(\mathfrak{s} \cap \Lambda)$ and $\overline{S} = \mathfrak{s}/\pi(\Lambda)$, where $\pi\colon \mathfrak{t} \to \mathfrak{s}$ is orthogonal projection.

1.1. Preliminaries

Let Φ be a root system on a real vector space V. Fix a set of simple roots Δ for Φ. Denote by $Q = Q(\Phi) \subset V^*$ the root lattice, $P = P(\Phi) \subset V^*$ the lattice of weights with basis $\{\varpi_\alpha\}_{\alpha\in\Delta}$. Further, we denote the inverse root system by $\Phi^\vee \subset V$. We have $Q^\vee = Q(\Phi^\vee) \subset V$ the coroot lattice. It is dual to P and the coweight lattice $P^\vee \subset V$ is dual to $Q(\Phi)$. We denote the Weyl group by $W(\Phi)$ or simply by W if Φ is clear from the context. We fix an inner product $\langle\,\cdot,\cdot\,\rangle$ on V invariant under the Weyl group and use it to identify V with V^*. We choose this inner product so that in each irreducible factor the shortest length of a coroot is $\sqrt{2}$. Coroots of this length are called *short coroots* whereas coroots of longer lengths are called *long coroots*. Roots dual to short coroots are thus *long roots* and roots dual to long coroots are *short roots*.

Associated to $a, b \in \Phi$ is the Cartan integer defined by

$$n(a,b) = \frac{2\langle a,b\rangle}{\langle b,b\rangle}. \tag{1.1}$$

The Cartan integers for pairs of elements in Δ determine and are determined by the Dynkin diagram $D(\Phi)$ together with the multiplicities and arrows of its bonds [**4**]. The set $\Delta^\vee \subseteq \Phi^\vee$ consists of the coroots $a^\vee$ inverse to each root $a \in \Delta$. The Cartan integers $n(a^\vee, b^\vee)$ for $a^\vee, b^\vee \in \Delta^\vee$ are described by a Dynkin diagram $D^\vee(\Phi)$, *the coroot diagram* for Φ. Its nodes are identified in the obvious way with the nodes of $D(\Phi)$. Its bonds, including the multiplicities, are exactly the same as the bonds in $D(\Phi)$, but the direction of every arrow is reversed.

Suppose that Φ is irreducible. Let d be the highest root of Φ with respect to Δ. Set $\tilde{a} = -d$ and let $\widetilde{\Delta} = \Delta \cup \{\tilde{a}\}$ be the *extended set of simple roots.* Let $C_0 \subseteq V$ be the positive Weyl chamber associated to Δ and let $A \subseteq C_0$ be the unique alcove in C_0 containing the origin. The walls of A are given by $\{a = 0\}_{a\in\Delta}$ and $\{\tilde{a} = -1\}$. Hence there is a natural bijection between $\widetilde{\Delta}$ and the walls of A. The set $\widetilde{\Delta}$ is the set of nodes for the extended Dynkin diagram $\widetilde{D}(\Phi)$. The Cartan integers $n(a,b)$ for $a, b \in \widetilde{\Delta}$ are recorded in the multiplicities of the bonds and the directions of the arrows of $\widetilde{D}(\Phi)$, by exactly the same rules as given in the case of $D(\Phi)$. (In the case of $\widetilde{A}_1$, we shall always make the convention that the two nodes are connected by two single bonds, so that the diagram is a cycle.) Dually, there is the extended coroot diagram $\widetilde{D}^\vee(\Phi)$ whose nodes are the coroots $\widetilde{\Delta}^\vee$ inverse to the roots in $\widetilde{\Delta}$. (N.B. The coroot inverse to the highest root is the highest short coroot, and is equal to the highest coroot if and only if Φ is simply laced.) As in the case of $D(\Phi)$, the diagram $\widetilde{D}^\vee(\Phi)$ is obtained from $\widetilde{D}(\Phi)$ by reversing the directions of all the arrows on the multiple bonds. In case $\Phi = \coprod_i \Phi_i$ is reducible and the Φ_i are the irreducible factors, we define the set of extended roots $\widetilde{\Delta}$ as $\coprod_i \widetilde{\Delta}_i$, and define the extended root and coroot diagrams as the disjoint union of the corresponding diagrams of the factors.

Still assuming that Φ is irreducible, there is a single linear relation among the roots of $\widetilde{\Delta}$, namely

$$\sum_{a\in\widetilde{\Delta}} h_a a = 0 \tag{1.2}$$

for positive integers h_a, with $h_{\tilde{a}} = 1$. The h_a will be called the *root integers*. The sum $\sum_{a\in\widetilde{\Delta}} h_a = h$ is the *Coxeter number* of the root system. Dually, there is a single linear relation between the coroots inverse to the roots in $\widetilde{\Delta}$. It takes the form

$$\sum_{a\in\widetilde{\Delta}} g_a a^\vee = 0 \tag{1.3}$$

where the g_a are all positive integers and the coefficient of the coroot $\tilde{a}^\vee$ is one. The integers g_a are called the *coroot integers* and the sum $g = \sum_{a\in\widetilde{\Delta}} g_a$ is called the *dual Coxeter number*. Since $\tilde{a}$ is a long root, it follows that $g_a|h_a$ for every $a \in \widetilde{\Delta}$, with equality exactly for the long roots in $\widetilde{\Delta}$. It will be convenient to view the coroot integers as defining a function $\mathbf{g}\colon \widetilde{\Delta} \to \mathbf{N}$ by the formula $\mathbf{g}(a) = g_a$.

For a root system Φ we define $\mathcal{C}\Phi$ to be the quotient $P^\vee(\Phi)/Q^\vee(\Phi)$. This is a finite abelian group. There is a representation of $\mathcal{C}\Phi$ as a group of affine isometries of V normalizing the alcove A, which will be described in more detail in Section 3.2. For each $c \in \mathcal{C}\Phi$, the differential w_c of its action on A is a linear map which is an element of $W(\Phi)$ normalizing $\widetilde{\Delta} \subset V^*$. Since the action of W preserves the Cartan integers, it follows that the resulting action of w_c on the nodes of $\widetilde{D}(\Phi)$ is a diagram automorphism. The action preserves the root integers h_a, in the sense that $h_{w_c\cdot a} = h_a$. Of course, w_c also induces a diagram automorphism of $\widetilde{D}^\vee(\Phi)$ preserving the coroot integers g_a. This defines a faithful representation of $\mathcal{C}\Phi$ as a group of diagram automorphisms of $\widetilde{D}(\Phi)$ and of $\widetilde{D}^\vee(\Phi)$.

Assume now that $\Phi = \Phi(T,G)$ is the root system of G with respect to T. It is a root system on the real vector space $V = i\mathfrak{t} \subset \mathfrak{t}\otimes\mathbf{C}$. By definition, the Dynkin diagram $D(G)$ of G is that of Φ and the extended Dynkin diagram $\widetilde{D}(G)$ is $\widetilde{D}(\Phi)$; similarly, we set $D^\vee(G) = D^\vee(\Phi)$ and $\widetilde{D}^\vee(G) = \widetilde{D}^\vee(\Phi)$.

The Weyl group $W(T,G) = N_G(T)/Z_G(T)$ is represented as a group of automorphisms of T and hence of $\mathfrak{t}$. On the other hand, $W(\Phi)$ is represented as a group of automorphisms of $i\mathfrak{t}$. These groups induce the same group of complex linear automorphisms of $\mathfrak{t}\otimes_{\mathbf{R}}\mathbf{C}$, and hence are canonically identified. Under this identification, for each root a of Φ the reflection r_a in $W(\Phi)$ is identified with the element in $W(T,G)$ which is reflection of $\mathfrak{t}$ in the plane $\mathrm{Ker}(a)\cap\mathfrak{t}$. (Here, by $\mathrm{Ker}(a)$ we implicitly mean the complexification of the kernel.)

Let $Q^\vee \subseteq P^\vee \subset i\mathfrak{t}$ be the coroot and coweight lattices, respectively, for Φ. The kernel of the exponential mapping $\mathfrak{t} \to T$ is $2\pi i Q^\vee$. Thus, the exponential map identifies $\mathfrak{t}/2\pi i Q^\vee$ with T and $\mathfrak{t}/2\pi i P^\vee$ with the maximal torus in the adjoint form of G. It follows that $\mathcal{C}\Phi = P^\vee/Q^\vee$ is identified with the center of G, $\mathcal{C}G$. Also, multiplication by $2\pi i$ identifies the affine Weyl group $W_{\mathrm{aff}}(\Phi)$ of Φ with the affine Weyl group $W_{\mathrm{aff}}(G)$ of G with respect to T, which by definition is the group of automorphisms of $\mathfrak{t}$ generated by the Weyl group $W(T,G)$ and translations by $\pi_1(T)$. It follows that the map $a \mapsto \exp(2\pi i a)$ identifies the alcove $A \subset i\mathfrak{t}$ with the quotient T/W and hence induces an identification of A with the space of conjugacy classes of elements in G. Under this identification the action of $\mathcal{C}\Phi$ on A, as defined above, agrees with the action of $\mathcal{C}G$ by multiplication on the space of conjugacy classes in G. Thus, this identification and the choice of the set of simple roots for Φ determine a homomorphism $\mathcal{C}G \to W(T,G)$ denoted by $c \mapsto w_c$. For any subgroup $\mathcal{C} \subset \mathcal{C}G$ we denote by $w_{\mathcal{C}}$ the image in $W(T,G)$ of $\mathcal{C}$ under this homomorphism. Let $S^{w_{\mathcal{C}}}$ be the subtorus of T whose Lie algebra is $\mathfrak{t}^{w_{\mathcal{C}}}$, the subspace of $\mathfrak{t}$ pointwise fixed under the action of $w_{\mathcal{C}}$. When $\mathcal{C}$ is the cyclic group generated by c we denote $\mathfrak{t}^{w_{\mathcal{C}}}$ by $\mathfrak{t}^{w_c}$ and $S^{w_{\mathcal{C}}}$ by S^{w_c}.

Through the identification of $\mathcal{C}\Phi$ and $\mathcal{C}G$, we have $\mathcal{C}G$ represented as a group of automorphisms of $\widetilde{\Delta}$ coming from a group of diagram automorphisms of $\widetilde{D}^\vee(G)$ preserving the coroot integers g_a. Let $\mathcal{C} \subset \mathcal{C}G$ be a subgroup. Denote by $\widetilde{\Delta}_{\mathcal{C}}$, resp. $\widetilde{\Delta}^\vee_{\mathcal{C}}$, the quotient of $\widetilde{\Delta}$, resp. $\widetilde{\Delta}^\vee$, under the action of $\mathcal{C}$. For each orbit $\overline{a} \in \widetilde{\Delta}_{\mathcal{C}}$ we define $g_{\overline{a}}$ to be $n_{\overline{a}} g_a$ where $n_{\overline{a}}$ is the cardinality of the orbit $\overline{a}$ and g_a is the coroot integer associated to any a in this orbit.

1.2. The case of commuting pairs in a simply connected group

In [**1**], the first author proved that, if x and y are commuting elements in a compact, simply connected group G, then there is a maximal torus $T \subseteq G$ containing both x and y. Furthermore, two pairs of elements in T are conjugate in G if and only if they are conjugate by an element of W. Thus, the moduli space of conjugacy classes of commuting pairs of elements in G is identified with $(T \times T)/W$. The torus T is the quotient of $\mathfrak{t}$ by the lattice $2\pi i Q^\vee$, where $Q^\vee$ is generated by $\widetilde{\Delta}^\vee$, and W is the group of isometries of $\mathfrak{t}$ generated by reflections in the hyperplanes through the origin defined by the elements of $\widetilde{\Delta}^\vee$. The group W acts on $\mathfrak{t}$ preserving the lattice $2\pi i Q^\vee$ and hence there is an induced W-action on T. This is the model result that we carry over into all the other cases we study.

1.3. c-pairs

Next, let us consider a compact, connected, simple, but not necessarily simply connected group K with G as simply connected covering. The first invariant of a K-bundle ξ over T^2 is the characteristic class $w(\xi) \in H^2(T^2;\pi_1(K)) = \pi_1(K)$. We identify $\pi_1(K)$ with a subgroup of the center $\mathcal{C}G$ of G, so that $w(\xi) \in \mathcal{C}G$. If ξ has a flat connection and if $x, y \in K$ are the holonomy images of the standard generators of the fundamental group of the two-torus, then for any lifts $\tilde{x} \in G$, resp. $\tilde{y} \in G$, of x, resp. y, we have $[\tilde{x},\tilde{y}] = w(\xi)$. Our classification results in this case are simplified by three assumptions:

(1) We fix the topological type of the bundle ξ, or equivalently the class $w(\xi) = c \in \mathcal{C}G$.
(2) We assume that ξ does not lift to a bundle over any non-trivial covering group of K.
(3) We classify flat connections on ξ up to restricted gauge equivalence, i.e., up to G-gauge equivalence. In this case, it turns out that restricted gauge equivalence is the same as K-equivalence.

Translating these conditions gives the following equivalent group theoretic problem.

Let G be simple, and let $c \in \mathcal{C}G$. A pair of elements (x,y) in G is said to be a c-pair if $[x,y] = c$. We classify c-pairs up to simultaneous conjugation by elements of G.

THEOREM 1.3.1. *Let G be simple, and let $c \in \mathcal{C}G$. Then the moduli space of c-pairs of elements in G, modulo simultaneous conjugation, is homeomorphic to* $(\overline{S}^{w_c} \times \overline{S}^{w_c})/W(S^{w_c}, G)$.

In a very closely related form, this theorem was first proved by Schweigert [**16**]. In Theorem 1.6.2 below, we shall describe $\overline{S}^{w_c}$ and $W(S^{w_c}, G)$ in terms of the extended coroot diagram of G and the action of c on this diagram.

1.4. Commuting triples

Next, we let G be simple and we turn to flat G-bundles over the 3-torus. The holonomy of such a connection around the standard basis of the fundamental group of the torus is a commuting triple (x, y, z) in G well-defined up to simultaneous conjugation. Let $\mathcal{T}_G$ denote the moduli space of conjugacy classes of commuting triples in G. In general, $\mathcal{T}_G$ has several components even though there is only one topological type for a G-bundle over T^3.

THEOREM 1.4.1. *Let G be simple. For any $k \geq 1$ dividing at least one of the coroot integers g_a we set $\widetilde{I}(k) = \{a \in \widetilde{\Delta} : k \nmid g_a\}$, and we let $S(k) \subseteq T$ be the subtorus with*

$$\mathrm{Lie}(S(k)) = \mathfrak{t}(k) = \bigcap_{a \in \widetilde{I}(k)} \mathrm{Ker}\, a.$$

Note that $\dim S(k)$ *is one less than the number of a such that $k | g_a$. Then:*

(1) *For each commuting triple (x, y, z) in G, there is a unique integer $k \geq 1$ dividing at least one of the coroot integers g_a such that $S(k)$ is conjugate to a maximal torus for $Z(x, y, z)$. The integer k is called the* order *of (x, y, z).*
(2) *The order is a conjugacy class invariant and defines a locally constant function on $\mathcal{T}_G$. We define the order of a component X of $\mathcal{T}_G$ to be the value of this function on X.*
(3) *If $k \geq 1$ divides at least one of the g_a, there are exactly $\varphi(k)$ components of $\mathcal{T}_G$ of order k, where φ is the Euler φ-function. Given a component X of $\mathcal{T}_G$, let $d_X = \frac{1}{3} \dim X + 1$. Then*
$$\sum_X d_X = g.$$
(4) *Each component of $\mathcal{T}_G$ of order k is homeomorphic to*
$$\left(\overline{S}(k) \times \overline{S}(k) \times \overline{S}(k)\right)/W(S(k), G).$$
(5) *Let π_k be orthogonal projection from $i\mathfrak{t}$ to $i\mathfrak{t}(k)$. For $a \in \widetilde{\Delta}$, $\pi_k(a^\vee)$ is non-zero if and only if $a \notin \widetilde{I}(k)$. Thus π_k determines an embedding of $\widetilde{\Delta}^\vee - \widetilde{I}^\vee(k)$ into $i\mathfrak{t}(k)$. Let $Q^\vee(k) \subset i\mathfrak{t}(k)$ be the lattice generated by this image. The torus $\overline{S}(k)$ is the quotient of $\mathfrak{t}(k)$ by the lattice $2\pi i Q^\vee(k)$. The group $W(S(k), G)$ is the group of isometries of $\mathfrak{t}(k)$ generated by reflections in the hyperplanes $\pi_k(a^\vee)^\perp$ for $a \in \widetilde{\Delta} - \widetilde{I}(k)$.*

Results along these lines have been obtained independently by Kac-Smilga [**11**].

In Section 1.7 we state a result which shows that the image of $\Delta^\vee - \widetilde{I}(k)^\vee$ in $i\mathfrak{t}(k)$ is the extended set of simple coroots of a root system $\Phi(\mathfrak{t}(k))$ on $i\mathfrak{t}(k)$ and explain how to derive its extended coroot diagram from the extended coroot diagram $\widetilde{D}^\vee(G)$ and the coroot integers g_a.

1.5. C-triples

Let K be a connected, simple group with simply connected covering G. Now we consider principal K-bundles with flat connection over the three-torus. Given a basis for the fundamental group of the 3-torus, a flat connection is determined by the holonomy image $x_1, x_2, x_3 \in K$ for the given basis of the fundamental group of the torus. These elements are defined up to simultaneous conjugation in K. Lifting these to elements $\tilde{x}_1, \tilde{x}_2, \tilde{x}_3$ in G, we have elements $c_{ij} = [\tilde{x}_i, \tilde{x}_j]$ in $\pi_1(K) \subseteq \mathcal{C}G$. These elements determine the topological type of the K-bundle. We record these elements by constructing 3×3 matrix $C = (c_{ij})$ with values in $\mathcal{C}G$. The entries of C satisfy: $c_{ii} = 1, c_{ij} = c_{ji}^{-1}$ for $i \neq j$. Such a matrix will be called *antisymmetric*. Let $\langle C \rangle$ be the subgroup of $\mathcal{C}G$ generated by the entries of C. A triple of elements (x_1, x_2, x_3) in G such that $c_{ij} = [x_i, x_j]$ is a *C-triple*. If $c_{12} = c$ and $c_{13} = c_{23} = 1$, we refer to (x_1, x_2, x_3) as a *c-triple*. As long as $\langle C \rangle$ is cyclic and generated by c, the moduli space of C-triples can be identified with the moduli space of c-triples. As in the case of the two-torus, we assume that the bundle does not lift to a proper covering group of K. This is equivalent to saying that the group $\langle C \rangle$ generated by the entries of C is all of $\pi_1(K)$. Also as before, we classify these bundles up to restricted gauge equivalence, i.e., up to conjugation by elements of G. In this case, this classification differs slightly from the classification up to conjugation by K. The classification of commuting triples in K follows easily from the results of this paper, but we shall not give the details here.

Denote by w_C the subgroup $w_{\langle C \rangle}$ of W which is the image of $\langle C \rangle$ under the homomorphism from $\mathcal{C}G$ to W described in Section 1.1. Let $\mathfrak{t}^{w_C}$ be the fixed subspace of the action of w_C on $\mathfrak{t}$, and let S^{w_C} be the subtorus of T whose Lie algebra is $\mathfrak{t}^{w_C}$. Finally, let $\widetilde{\Delta}_C$, resp. $\widetilde{\Delta}_C^\vee$, be the quotients of $\widetilde{\Delta}$, resp. $\widetilde{\Delta}^\vee$, by the action of w_C.

THEOREM 1.5.1. *Let G be simple and let C be an antisymmetric 3×3-matrix with values in $\mathcal{C}G$. Let $\mathcal{T}_G(C)$ be the moduli space of conjugacy classes of C-triples in G. Let $\overline{\mathbf{g}} \colon \widetilde{\Delta}_C \to \mathbf{N}$ be the function defined by $\overline{\mathbf{g}}(\overline{a}) = g_{\overline{a}}$. For any $k \geq 1$ dividing at least one of the $g_{\overline{a}}$, let $\widetilde{I}_C(k)$ be the set of $\overline{a} \in \widetilde{\Delta}_C$ such that $k \nmid g_{\overline{a}}$ and let $S^{w_C}(\overline{\mathbf{g}}, k) \subseteq S^{w_C}$ be the subtorus whose Lie algebra is*

$$\mathfrak{t}^{w_C}(\overline{\mathbf{g}}, k) = \bigcap_{\overline{a} \in \widetilde{I}_C(k)} \operatorname{Ker} \overline{a} \subseteq \mathfrak{t}^{w_C}.$$

Let π_k^C be orthogonal projection from $\mathfrak{t}$ to $\mathfrak{t}^{w_C}(\overline{\mathbf{g}}, k)$. Then:

(1) *For every C-triple (x, y, z), there is an integer $k \geq 1$ dividing at least one of the $g_{\overline{a}}$, called the* order *of (x, y, z), such that $S^{w_C}(\overline{\mathbf{g}}, k)$ is conjugate to a maximal torus for $Z(x, y, z)$. The dimension of $S^{w_C}(\overline{\mathbf{g}}, k)$ is one less than the number of $\overline{a}$ such that $k | g_{\overline{a}}$.*
(2) *The order is a conjugacy class invariant and defines a locally constant function on $\mathcal{T}_G(C)$. We define the order of a component X of $\mathcal{T}_G(C)$ to be the value of this function on X.*
(3) *For any $k \geq 1$ dividing at least one of the $g_{\overline{a}}$, there are exactly $\varphi(k)$ components of $\mathcal{T}_G(C)$ of order k. Given a component X of $\mathcal{T}_G(C)$, let $d_X = \frac{1}{3} \dim X + 1$. Then*

$$\sum_X d_X = g.$$

(4) *In the case where $\langle C\rangle$ is cyclic, i.e. the case of c-triples, each component of $\mathcal{T}_G(C)$ of order k is homeomorphic to either*

$$\left(\overline{S}\times\overline{S}\times\overline{S}\right)/W(S,G) \quad or \quad \left(\overline{S}\times\overline{S}\times S\right)/W(S,G),$$

where $S = S^{w_c}(\overline{\mathfrak{g}},k)$. We will describe the possibilities precisely in Chapter 8. In case $\langle C\rangle$ is not cyclic, each component of the moduli space is homeomorphic to

$$((S\times S\times S)/F)/W(S,G)),$$

where $S = S^{w_C}(\overline{\mathfrak{g}},k)$ and F is a finite subgroup of $S\times S\times S$ which will be described explicitly in Chapter 11.

(5) *For $a\in\widetilde{\Delta}$, $\pi_k^C(a^\vee)$ depends only on $\overline{a}\in\widetilde{\Delta}_C$ and is nonzero if and only if $\overline{a}\notin\widetilde{I}_C(k)$. Thus π_k^C determines an embedding of $\widetilde{\Delta}_C^\vee-\widetilde{I}_C^\vee(k)$ into $i\mathfrak{t}^{w_C}(\overline{\mathfrak{g}},k)$. Let $Q_C^\vee(k)\subset i\mathfrak{t}^{w_C}(\overline{\mathfrak{g}},k)$ be the lattice spanned by the image. The torus $\overline{S}^{w_C}(\overline{\mathfrak{g}},k)$ is the quotient of $\mathfrak{t}^{w_C}(\overline{\mathfrak{g}},k)$ by the lattice $2\pi i Q_C^\vee(k)$. The group $W(S^{w_C}(\overline{\mathfrak{g}},k),G)$ is the group of isometries of $\mathfrak{t}^{w_C}(\overline{\mathfrak{g}},k)$ generated by reflections in the hyperplanes $\pi_k^C(a^\vee)^\perp$ for $\overline{a}\in\widetilde{\Delta}_C-\widetilde{I}_C(k)$.*

In Section 1.7 we state a result which shows that the image of $\widetilde{\Delta}_C^\vee-\widetilde{I}_C^\vee(k)$ in $i\mathfrak{t}^{w_C}(\overline{\mathfrak{g}},k)$ is the extended set of simple coroots of a root system, and which describes its extended coroot diagram in terms of the extended coroot diagram $\widetilde{D}^\vee(G)$, the action of $\langle C\rangle$ on this diagram, and the coroot integers g_a.

1.6. Quotients of diagram automorphisms

Let $\widetilde{D}$ be an extended coroot diagram. A diagram automorphism $\sigma\colon\widetilde{D}\to\widetilde{D}$ automatically preserves the coroot integers. Suppose that $\mathcal{C}$ is a group of diagram automorphisms of $\widetilde{D}$.

DEFINITION 1.6.1. We form the quotient diagram $\widetilde{D}/\mathcal{C}$ as follows. The nodes of $\widetilde{D}/\mathcal{C}$ are the $\mathcal{C}$-orbits of nodes of $\widetilde{D}$. The orbit of the node v is denoted $\overline{v}$.

There is one case that we handle directly. If $\widetilde{D}$ is $\widetilde{A}_n$ and $\mathcal{C}$ acts transitively on the set of nodes of $\widetilde{D}$, then the quotient has one element.

Ruling out this degenerate case, there are two types of $\mathcal{C}$-orbits on an extended diagram $\widetilde{D}$: an orbit consisting of nodes, no two of which are connected by a bond, and an orbit consisting of subdiagrams of type A_2, no two of which are connected by a bond. The first type of orbit is called *ordinary* and the second *exceptional*. Let $\epsilon(\overline{v})=1$ if $\overline{v}$ is an ordinary orbit and let $\epsilon(\overline{v})=2$ if $\overline{v}$ is exceptional. We describe the bonds of $\widetilde{D}/\mathcal{C}$, their multiplicities and their arrows by giving the Cartan integers $n(\overline{u},\overline{v})$. If $\overline{u}$ and $\overline{v}$ are distinct orbits such that $n(u_i,v_j)=0$ for all $u_i\in\overline{u}$ and $v_j\in\overline{v}$, then $n(\overline{u},\overline{v})=0$ and there is no bond connecting $\overline{u}$ and $\overline{v}$ in the quotient diagram. If there are nodes $u\in\overline{u}$ and $v\in\overline{v}$ which are connected by a bond, then either $\mathrm{Stab}(u)\subseteq\mathrm{Stab}(v)$ or $\mathrm{Stab}(v)\subseteq\mathrm{Stab}(u)$. In the first case we define

$$n(\overline{u},\overline{v})=\epsilon(\overline{v})n(u,v),$$

and in the second case we define

$$n(\overline{u},\overline{v})=\epsilon(\overline{v})\frac{n_{\overline{v}}}{n_{\overline{u}}}n(u,v),$$

where as before $n_{\overline{u}}$ is the cardinality of the orbit $\overline{u}$. It is easy to see that these numbers are well-defined integers and satisfy: $n(\overline{u},\overline{v})=0$ implies $n(\overline{v},\overline{u})=0$;

otherwise $n(\overline{u}, \overline{v}) < 0$ for $\overline{u} \neq \overline{v}$. Thus, these numbers determine the bonds of the quotient diagram $\widetilde{D}/\mathcal{C}$, their multiplicities and the directions of their arrows.

We have the set $\{g_v\}$ of coroot integers on the original diagram. We define integers $g_{\overline{v}}$ on the nodes of $\widetilde{D}/\mathcal{C}$ as follows. In all cases, including the degenerate ones, the induced coroot integers $g_{\overline{v}}$ on the quotient diagram are defined by $g_{\overline{v}} = \sum_v g_v$ where v ranges over the nodes in the orbit $\overline{v}$. Since diagram automorphisms preserve the integers g_v, we have $g_{\overline{v}} = n_{\overline{v}} g_v$ for any $v \in \overline{v}$.

Here is the theorem which describes the fixed tori and their Weyl groups in terms of diagram automorphisms.

THEOREM 1.6.2. *Let Φ be a irreducible root system on a real vector space V, and let $\mathcal{C}$ be a subgroup of $\mathcal{C}\Phi$. Let $\pi\colon V \to V^{w_\mathcal{C}}$ be orthogonal projection. Then:*

(1) *Restriction of π to $\widetilde{\Delta}^\vee$ induces an embedding of $\widetilde{\Delta}^\vee_\mathcal{C}$ in $V^{w_\mathcal{C}}$. This embedding identifies the set of nodes $\widetilde{\Delta}^\vee$ of $\widetilde{D}^\vee/\mathcal{C}$ with an extended set of simple coroots for a root system $\Phi(w_\mathcal{C})$.*
(2) *$\widetilde{D}^\vee/\mathcal{C}$ is the extended coroot diagram of $\Phi(w_\mathcal{C})$.*
(3) *The coroot lattice of $\Phi(w_\mathcal{C})$ is equal to $\pi(Q^\vee)$, and the group $W(S^{w_\mathcal{C}}, \Phi)$ is the Weyl group of $\Phi(w_\mathcal{C})$.*

This result together with Theorem 1.3.1 yields an explicit description of the moduli space of c-pairs in terms of the extended coroot diagram of G and the action of w_c on this diagram.

1.7. Description of $\overline{S}(k)$ and $\overline{S}^{w_\mathcal{C}}(\overline{\mathbf{g}}, k)$

Let $\widetilde{D}$ be a connected extended coroot diagram, whose set of nodes is $\widetilde{\Delta}^\vee$, and let $\mathbf{n}\colon \widetilde{\Delta} \to \mathbf{N}$ be a function of the form $n_0\mathbf{g}$ for some positive integer n_0. Given an integer $k \geq 1$ dividing at least one of the $\mathbf{n}(a)$, let $\widetilde{I}(\mathbf{n}, k) = \{a \in \widetilde{\Delta} : k \nmid \mathbf{n}(a)\}$. We let $\widetilde{D}'(\mathbf{n}, k)$ be the largest subdiagram of $\widetilde{D}$ having $\widetilde{I}^\vee(\mathbf{n}, k)$ as its set of nodes. Fix a length function $\ell\colon A \to \mathbf{R}^+$ such that if $v, v' \in \widetilde{\Delta}^\vee$ are connected by a bond of $\widetilde{D}$ and $\ell(v) \geq \ell(v')$ then the multiplicity of the bond between v and v' is $\ell(v)^2/\ell(v')^2$ and the arrow (if the multiplicity is not one) points toward v'.

PROPOSITION 1.7.1. *Let $v \in \widetilde{\Delta}^\vee - \widetilde{I}^\vee(\mathbf{n}, k)$. Then exactly one of the following holds.*

(Type ∞) *v is the only element of $\widetilde{\Delta}^\vee - \widetilde{I}^\vee(\mathbf{n}, k)$.*
(Type 1) *The node v is not connected by a bond of $\widetilde{D}$ to a node of $\widetilde{D}'(\mathbf{n}, k)$, but is not of Type ∞.*
(Type 2)(i) *The node v is connected by bonds of $\widetilde{D}$ to exactly two components of $\widetilde{D}'(\mathbf{n}, k)$, each of which is of type A_1, with nodes which we denote by v_1 and v_2, say, and $\ell(v) \leq \min(\ell(v_1), \ell(v_2))$.*
(Type 2)(ii) *The node v is connected by bonds of $\widetilde{D}$ to exactly one node v_1 of $\widetilde{D}'(\mathbf{n}, k)$, and $\ell(v) < \ell(v_1)$.*
(Type 3) *The node v is connected by bonds of $\widetilde{D}$ to exactly two components of $\widetilde{D}'(\mathbf{n}, k)$, each of which is of type A_2.*
(Type 4) *The node v is connected by bonds of $\widetilde{D}$ to exactly two components, and exactly one of the following holds:*
  (i) *Both of these components are of type A_1, with nodes denoted by v_1 and v_2, say, and $\ell(v) > \min(\ell(v_1), \ell(v_2))$;*

(ii) *Each component is of type* A_3*;*
(iii) *One component is of type* A_3 *and the other is of type* A_1*.*

DEFINITION 1.7.2. Define a new length function $\ell_k \colon \widetilde{\Delta}^\vee - \widetilde{I}^\vee(\mathbf{n}, k) \to \mathbf{R}^+$ by setting $\ell_k(v) = \ell(v)/\sqrt{r}$ if v is of Type r according to the above proposition. We will see later that, if $\ell_k(v) \geq \ell_k(w)$, then in fact $\ell_k(v)^2/\ell_k(w)^2 \in \mathbf{Z}$.

Let $\widetilde{D}(\mathbf{n}, k)$ be the unique diagram with nodes $\widetilde{\Delta}^\vee - \widetilde{I}^\vee(\mathbf{n}, k)$ satisfying the following three conditions:

(1) Nodes $v, v' \in \widetilde{\Delta}^\vee - \widetilde{I}^\vee(\mathbf{n}, k)$ are connected by a bond in $\widetilde{D}(\mathbf{n}, k)$ if and only if either v and v' are connected by a bond in $\widetilde{D}$ or v and v' are connected by bonds of $\widetilde{D}$ to the same component of $\widetilde{D}'(\mathbf{n}, k)$.
(2) Suppose that v and v' are distinct nodes of $\widetilde{\Delta}^\vee - \widetilde{I}^\vee(\mathbf{n}, k)$ which are connected by a bond in $\widetilde{D}(\mathbf{n}, k)$ and suppose that neither of these nodes is connected by a bond of $\widetilde{D}(\mathbf{n}, k)$ to any other node of $\widetilde{\Delta}^\vee - \widetilde{I}^\vee(\mathbf{n}, k)$. If $\ell_k(v) = \ell_k(v')$, then the subdiagram of $\widetilde{D}(\mathbf{n}, k)$ spanned by v and v' is of type $\widetilde{A}_1$.
(3) Suppose that v and v' are distinct nodes of $\widetilde{\Delta}^\vee - \widetilde{I}^\vee(\mathbf{n}, k)$ connected by a bond of $\widetilde{D}(\mathbf{n}, k)$ and suppose that v and v' are not as in the previous case. Lastly, suppose that $\ell_k(v) \geq \ell_k(v')$. Then the multiplicity of the bond in $\widetilde{D}(\mathbf{n}, k)$ connecting v to v' is $\ell_k(v)^2/\ell_k(v')^2$ and the arrow (if this multiplicity is greater than one) points toward v'.

PROPOSITION 1.7.3. *The diagram* $\widetilde{D}(\mathbf{n}, k)$ *defined above is the coroot diagram of a (possibly non-reduced) root system.*

In particular, taking the quotient diagram $\widetilde{D}^\vee/\langle C\rangle$ as our extended coroot diagram and the function $\overline{\mathbf{g}}$ defined in Theorem 1.5.1 as $\mathbf{n}$ defines an extended coroot diagram $(\widetilde{D}^\vee/\langle C\rangle)(\overline{\mathbf{g}}, k)$. The next theorem shows that this diagram is in fact the extended coroot of a root system on $i\mathfrak{t}^{w_C}(\overline{\mathbf{g}}, k)$.

THEOREM 1.7.4. *Let* G *be simple, let* C *be an antisymmetric* 3×3 *matrix with values in* $\mathcal{C}G$*, and let* $k \geq 1$ *be an integer dividing at least one of the* $g_{\overline{a}}$*. Orthogonal projection from* $i\mathfrak{t}$ *to* $i\mathfrak{t}^{w_C}(\overline{\mathbf{g}}, k)$ *induces an embedding of* $\widetilde{\Delta}^\vee_C - \widetilde{I}^\vee_C(k)$ *in* $i\mathfrak{t}^{w_C}(\overline{\mathbf{g}}, k)$*. This embedding identifies the nodes* $\widetilde{\Delta}^\vee_C - \widetilde{I}^\vee_C(k)$ *of* $(\widetilde{D}^\vee/\langle C\rangle)(\overline{\mathbf{g}}, k)$ *with an extended set of simple coroots of a root system* $\Phi(w_C, k)$ *on* $i\mathfrak{t}^{w_C}(\overline{\mathbf{g}}, k)$*. The diagram* $(\widetilde{D}^\vee/\langle C\rangle)(\overline{\mathbf{g}}, k)$ *is the extended coroot diagram of* $\Phi(w_C, k)$*.*

This theorem, together with Theorem 1.4.1 and Theorem 1.5.1 lead to an explicit description of the components of the moduli spaces of commutative triples and C-triples in terms of the diagrams $\widetilde{D}^\vee(G)$ and $\widetilde{D}^\vee(G)/\langle C\rangle$ and the integers $g_{\overline{a}}$.

1.8. Chern-Simons invariants and Witten's "Clockwise Symmetry Conjecture"

Assume that G is simple. A connection on a principal G-bundle ξ over a three-manifold M has a Chern-Simons invariant, see [**5**], which measures its difference from a trivial connection. This invariant is well-defined modulo $\mathbf{Z}$ on isomorphism classes of connections. The Chern-Simons function is constant on continuous paths of flat connections, so that we can view it as a function from the components of the moduli space of gauge equivalence classes of flat connections on ξ to $\mathbf{R}/\mathbf{Z}$. In case

M is the three-torus, this moduli space is identified with the space of conjugacy classes of commuting triples in G, so that the Chern-Simons invariant defines a function from the set of components of $\mathcal{T}_G$ to $\mathbf{R}/\mathbf{Z}$.

Let us consider now principal K-bundles, where K is compact, connected and simple. Let G be the universal covering group of K. The topological type of a K-bundle ξ over T^3 is determined by its characteristic class $w(\xi) \in H^2(T^3; \pi_1(K))$. We construct an antisymmetric matrix $C(\xi) = (c_{ij}(\xi))$ where $c_{ij}(\xi)$ is the value of $w(\xi)$ on the coordinate two-torus T_{ij} in the $(ij)^{\text{th}}$-coordinate directions.

In this case, there is only a relative Chern-Simons invariant which is well-defined modulo $(1/n)\mathbf{Z}$ for some integer $n \geq 1$ depending on K. To get a Chern-Simons invariant well-defined modulo $\mathbf{Z}$ we need to consider enhanced K-bundles over a three-manifold M, where by an enhanced K-bundle Ξ we mean an underlying K-bundle ξ together with a lifting to G of the structure group of ξ over the one-skeleton of M. We define $C(\Xi) = C(\xi)$. Given an antisymmetric matrix C with coefficients in $\pi_1(K)$, there is, up to isomorphism, a unique enhanced K-bundle Ξ with $C(\Xi) = C$. Under automorphisms of Ξ, the Chern-Simons invariant is well-defined modulo $\mathbf{Z}$.

Let A be a flat connection on a K-bundle ξ over T^3. The holonomy of A is identified with a conjugacy class of commuting triples in K. Given an enhanced structure Ξ on ξ, by a flat connection on Ξ we simply mean a flat connection A on ξ. The connection A then lifts over the 1-skeleton to a flat G-connection, and this defines a lifting of the holonomy of A to a conjugacy class of $C(\xi)$-triples in G, called the *G-holonomy* of A on Ξ. The G-holonomy determines a bijection between the set of isomorphism classes of flat connections on enhanced K-bundles Ξ over T^3 with $C(\Xi) = C$ and the moduli space $\mathcal{T}_G(C)$ of conjugacy classes of C-triples in G.

THEOREM 1.8.1. *Let G be simple, let C be an antisymmetric matrix with entries in $\mathcal{C}G$, and let $K = G/\langle C\rangle$. Let X_1 be the unique component of order* 1 *of the moduli space of C-triples. Let Ξ be an enhanced K-bundle with $C(\Xi) = C$. Let Γ_1 be a flat connection on Ξ whose G-holonomy is a conjugacy class in X_1. Let A be a flat connection on Ξ. Then* $\mathrm{CS}_{\Gamma_1}(A) \bmod \mathbf{Z}$ *is independent of the choice of* Γ_1. *Let* $\mathrm{CS}(A)$ *be the class of* $\mathrm{CS}_{\Gamma_1}(A) \bmod \mathbf{Z}$.

(1) *The function* $\mathrm{CS}(A)$ *is a well-defined function on the set of isomorphism classes of flat connections on Ξ, or equivalently on $\mathcal{T}_G(C)$, to* $\mathbf{R}/\mathbf{Z}$.
(2) *Viewing* CS *as a function on $\mathcal{T}_G(C)$, it is constant on components.*
(3) *The function* CS *induces a bijection from the set of components of $\mathcal{T}_G(C)$ of order k to the set of points of order k in* $\mathbf{R}/\mathbf{Z}$.

There is a refinement of this theorem, leading to a surprising symmetry involving the Chern-Simons invariants and the dimensions of the components of $\mathcal{T}_G(C)$. It was first discovered by Witten from considerations in quantum field theory.

THEOREM 1.8.2. *Let G be simple and let C be an antisymmetric matrix with entries in $\mathcal{C}G$. Let g be the dual Coxeter number of G. For each component X of $\mathcal{T}_G(C)$, let d_X be $\frac{1}{3}\dim(X) + 1$, and let $\mathrm{CS}(X)$ be the value of the Chern-Simons functional on this component. Let $J(X) \subseteq \mathbf{R}/\mathbf{Z}$ be a subset of d_X equally spaced points centered at $\mathrm{CS}(X)$ with spacing $1/g$. Then the $J(X)$ are disjoint subsets of*

R/**Z**. *Let*

$$J = \bigcup_X J(X) \subset \frac{1}{2g}\mathbf{Z}/\mathbf{Z}.$$

Then J consists of g points invariant under translation by $1/g$. Thus, identifying **R**/**Z** *with the unit circle, the set J is invariant under by a rotation through angle $2\pi/g$.*

1.9. Outline of the paper

We begin in Chapter 2 with a discussion of almost commuting n-tuples in a compact group K. We define the rank of such an n-tuple to be the dimension of a maximal torus of the centralizer of the n-tuple. We show a general finiteness statement: for a compact group K, there are only finitely many conjugacy classes of rank zero almost commuting triples in K. More generally, given a torus S in K, we give an explicit description of the subspace of the moduli space of conjugacy classes of almost commuting n-tuples $\mathbf{x}$ such that S is conjugate to a maximal torus of the centralizer of $\mathbf{x}$. Each such component is a quotient of the product of n copies of S by a finite group. The problem now is to determine the possible tori and finite groups which arise in the cases we study in more detail.

Chapter 3 contains various characterizations of groups of type A. These results are applied to deduce the following property of the coroot integers g_a associated to the simple roots a: The subdiagram D_k of the extended Dynkin diagram of G consisting of all nodes corresponding to roots a for which g_a is not divisible by a fixed positive integer k is a disjoint union of diagrams of A_{n_i}-type, for integers n_i with the property that $(n_i+1)|k$. Furthermore, if I_i denotes the set of simple roots corresponding to the nodes of one of the components of D_k, then the exponential of $2\pi i \sum_{a\in I_i}(g_a/k)a^\vee$ lies in the center of the corresponding simple subgroup of G and generates the center. This result implies that the set of integers which occur as g_a form an interval $[1, N]$ and that, for the Euler φ-function, one has $\varphi(N) = 2$, so that $N \in \{1, 2, 3, 4, 6\}$. These properties, as well as the characterization of groups of A-type, are important in the study of c-pairs and c-triples as well as for the numerology of clockwise symmetry.

Chapter 4 takes up the case of c-pairs and contains a proof of the first item in Theorem 1.3.1. Following the pattern laid down in Chapter 2, we first consider rank zero c-pairs and then pass to the general case of higher rank.

Chapter 5 considers the case of commuting triples in G, once again taking up the case of commuting triples of rank zero first. This chapter contains a proof of the first four parts of Theorem 1.4.1. While this chapter will be for the most part subsumed in Chapter 8, it seemed worthwhile to give the reasonably straightforward arguments needed to handle this case.

Chapters 6 and 7 are preparatory for the study of C-triples. They examine questions which, to us, are interesting in their own right. Chapter 6 considers a group τ of affine automorphisms of a vector space which normalizes an alcove of a root system on that vector space. Such automorphisms are equivalent to diagram automorphisms of the extended Dynkin diagram of the root system. We show that the walls of the original root system divide the fixed point set of τ into alcoves and that these are the alcoves of a Coxeter group acting on the fixed point set. Thus, there is a reduced root system on the underlying vector space of the fixed point set whose alcove structure is identified with the given alcove structure on the affine subspace fixed by τ. We use this to study two closely related root systems on the linear subspace fixed by the differential of τ. One root system, the restricted root system, consists of the non-trivial restrictions of roots of the original system to the fixed subspace. The other, the projection system, is the inverse root system to the root system obtained by taking the nonzero orthogonal projections of the coroots of the original system. These root systems are not in general equal, nor are they always inverse systems, nor is either equal in general to the root system on the fixed point set of the affine automorphism. It is also not true in general that these roots systems are reduced. Nevertheless, all three of them have the same Weyl groups, and hence the same set of roots up to positive multiples. We apply these results to the torus S^{w_c} fixed by the w_c-action on T. The main result is the completion of the proof of Parts 2 and 3 of Theorem 1.3.1 describing the Weyl group of this torus and the fundamental group of $\overline{S}^{w_c}$. We conclude by proving Theorem 1.6.2.

Chapter 7 is concerned with the centralizer $Z(x, y)$ of a c-pair (x, y). We describe the root system corresponding to the Lie algebra of $Z(x, y)$, the fundamental group, and the component group of $Z(x, y)$. We study $Z(x, y)$ by viewing conjugation by y as an automorphism of the compact group $Z(x)$. It is natural to consider the most general problem along these lines: let H be an arbitrary compact connected group and let σ be an automorphism of H. We study the component group of the fixed subgroup H^σ of σ. This description of the component group of H^σ contains, in particular, a generalization of the result of the first author in [**1**], that H^σ is connected if H is simply connected. Related results were proved by Steinberg in [**18**]. We also describe the Lie algebra of H^σ, generalizing work of Kac [**9**] in case σ has finite order.

With the preliminary work about centralizers of c-pairs established, we turn in Chapter 8 to c-triples, for c non-trivial. Following the general pattern, we first consider the case of c-triples of rank zero. It turns out there there is a finite and short list of simple groups that have rank zero c-triples. We then go on to establish Parts 1 through 4 of Theorem 1.5.1 in the case when $\langle C\rangle$ is cyclic.

In Chapter 9 we turn to the tori $\overline{S}(k)$ and $\overline{S}^{w_c}(\mathbf{g}, k)$ and compute their Weyl groups and fundamental groups. The result is a proof of Theorem 1.7.4, of Part 3 of Theorem 1.4.1 and of Part 3 of Theorem 1.5.1 in the case when C is cyclic. Once again, we find a related Coxeter group by arguments which are formally similar to those used to study the fixed subspace of an affine automorphism.

Chapter 10 considers the Chern-Simons invariant of a flat connection with holonomy a given c-triple. This invariant is identified with an invariant of the c-triple defined using the Weyl invariant inner product on the Cartan subalgebra of G. We then prove Theorem 1.8.1. Lastly, we establish Witten's clockwise symmetry statement, Theorem 1.8.2, in the case of c-triples.

In Chapter 11, we consider the case when the group $\langle C\rangle$ is non-cyclic, establishing both Theorems 1.5.1 and 1.8.2 in these cases by explicit computation.

At the end of the paper, we give a list of the possible coroot diagrams and quotient coroot diagrams, and give tables summarizing the tori and root systems we have defined, as well as other relevant information.

A guiding principle of this paper has been to avoid classification and case-by-case checking wherever possible. We have preferred to give more general, conceptual arguments, even at the cost of increasing the length.

1.10. History

Questions related to the ones considered here have a long history. One motivation was to understand the experimental connection between the torsion primes of G, i.e. the primes p for which there is p-torsion in the integral homology of G, and those primes p for which G has an elementary abelian p-group of rank three contained in a torus. The first author and J-P. Serre proved that if p is not a torsion prime, then every elementary p-group in G is contained in a torus. The converse was checked, case-by-case, in [**1**] and in the earlier references cited in that paper. From the point of view of this paper, the subgroups constructed in [**1**] are simply the commuting triples of rank zero. It was also checked that a torsion prime divides one of the root integers h_a, but the converse does not quite hold. R. Steinberg pointed out later that the torsion primes listed in [**1**] are exactly the prime divisors of the coroot integers g_a. In [**19**], he shows that the prime p divides one of the g_a if and only if G contains an elementary abelian p-subgroup not contained in a torus. The methods of the present paper are, in part, quite similar to those of [**19**]. More recently, Griess [**8**] classified the possibilities for elementary abelian p-groups in G, apparently by completely different methods from those of this paper. The last two authors of the present paper were led to this circle of problems from another direction, the study of holomorphic principal G-bundles over an elliptic curve. In this case, if the bundle is flat, then its holonomy defines a c-pair up to conjugation. The singularities of the moduli space of such bundles are then closely related to the component groups of centralizers of c-pairs, and thus ultimately to c-triples. We should also add that Kac-Smilga in [**11**] as well as Keurentjes [**12, 13**] have independently established results overlapping significantly with ours in the case of commuting triples. Their approach seems very similar to ours.

It is natural to ask about the moduli space of almost commuting N-tuples for $N > 3$. We believe that the methods developed here can attack this question as well. However, and perhaps not surprisingly, there seem to be very few essentially new cases. Along these lines, in [**11**], Kac and Smilga have classified a very special type of commuting N-tuple which they call minimal.

The impetus for the study we carry out here of commuting triples and c-triples was questions, conjectures, and statements of Witten about the nature of the moduli space, and especially the number and structure of its components. He was led to these questions by studying the quantum field theory of gauge theories over the three torus, in particular the R-symmetries of these theories. With this heuristic guide, he conjectured the clockwise-symmetry statement and checked it in many cases. It is our pleasure to thank Witten for pointing out these questions to us, and for many stimulating discussions on these and other related matters.

CHAPTER 2

Almost commuting N-tuples

An ordered N-tuple $\mathbf{x} = (x_1, \ldots, x_N)$ of elements in G is *almost commuting* if $[x_i, x_j] \in \mathcal{C}G$ for every $1 \le i, j \le N$. Notice that $\mathbf{x}$ is almost commuting if and only if the image ordered N-tuple $\overline{\mathbf{x}} = (\overline{x}_1, \ldots, \overline{x}_N)$ in the adjoint form of G is a commuting ordered N-tuple. Given an almost commuting N-tuple $\mathbf{x}$ let $c_{ij} = [x_i, x_j]$ and let C be the $N \times N$ matrix $C = (c_{ij})$ of elements in $\mathcal{C}G$. The matrix C is antisymmetric in the sense that $c_{ii} = 1$ and $c_{ij} = c_{ji}^{-1}$ for all i, j. We say that $\mathbf{x}$ is an *ordered N-tuple of C-type*.

Clearly, the space $\widetilde{\mathcal{M}}_G(C)$ of all ordered N-tuples in G of C-type is identified with a closed subspace of $\prod_{i=1}^N G$ and thus is a compact Hausdorff space. The compact group G acts on $\prod_{i=1}^N G$ by simultaneous conjugation normalizing the subspace $\widetilde{\mathcal{M}}_G(C)$. Hence, the quotient $\mathcal{M}_G(C) = \widetilde{\mathcal{M}}_G(C)/G$ is a compact Hausdorff space. It is the space of conjugacy classes of ordered N-tuples in G of C-type. When G is clear from context, we shall denote this space by $\mathcal{M}(C)$. Our goal in this chapter is to prove a very general qualitative result concerning $\mathcal{M}(C)$. We shall show that its connected components are homeomorphic to quotients of products of subtori of T by finite groups.

2.1. An invariant for almost commuting N-tuples

We define the *rank* of an ordered N-tuple $\mathbf{x}$, denoted $\mathrm{rk}(\mathbf{x})$, to be the rank of $Z(\mathbf{x})$, the centralizer of $\mathbf{x}$ in G. Notice that $\mathbf{x}$ is of rank zero if and only if $Z(\mathbf{x})$ is a finite group. There is a related but finer invariant of $\mathbf{x}$ derived from any maximal torus of $Z(\mathbf{x})$ which we shall now describe.

For any subset $I \subseteq \Delta$ let $\mathfrak{t}_I = \bigcap_{a \in I} \mathrm{Ker}\, a \subseteq \mathfrak{t}$, and let S_I be the subtorus of T with Lie algebra $\mathfrak{t}_I$. We denote by L_I the derived group of $Z(S_I)$:

$$L_I = DZ(S_I).$$

Since $Z(S_I)$ is the centralizer of a torus, it is connected, and thus L_I is also connected.

LEMMA 2.1.1. (1) *L_I is non-trivial if and only if $I \neq \Delta$.*

(2) *If L_I is non-trivial, then I forms a set of simple roots for L_I.*

(3) *L_I is simply connected.*

(4) *S_I is the component of the identity of the center of its centralizer.*

(5) *Conversely, if S is a torus in G and is equal to the component of the identity of the center of its centralizer, then there is a subset $I \subseteq \Delta$ such that S is conjugate to S_I.*

PROOF. All of these elementary facts are proved in [**4**]. □

LEMMA 2.1.2. *Let* $\mathbf{x}$ *be an ordered* N*-tuple and let* S *be a maximal torus of* $Z(\mathbf{x})$. *There is a unique subset* $I \subseteq \Delta$ *such that* S *is conjugate to* S_I.

PROOF. Let S' be the component of the identity of the center of $Z(S)$. Clearly, since S is contained in the center of $Z(S)$, $S \subseteq S'$. Since $\mathbf{x} \subseteq Z(S)$, we have $S' \subseteq Z(\mathbf{x})$. Since S is a maximal torus of $Z(\mathbf{x})$, this implies that $S' = S$. Since S is the identity component of its centralizer, S is conjugate to S_I for some $I \subseteq \Delta$. Uniqueness of I is clear. □

The subset $I \subseteq \Delta$ given in the last lemma is an invariant of the ordered N-tuple $\mathbf{x}$ and is denoted by $I(\mathbf{x})$. The cardinality of $\Delta - I(\mathbf{x})$ is the rank of $\mathbf{x}$.

LEMMA 2.1.3. *The torus* S *is a maximal torus of* G *if and only if the components* x_i *are contained in* S, *and in this case* $C = \mathrm{Id}$.

PROOF. If the x_i are all contained in the torus S, then they are mutually commuting. Moreover, if T is a maximal torus containing S, then $T \subseteq Z(\mathbf{x})$. Thus, since S is a maximal torus of $Z(\mathbf{x})$, $S = T$. Conversely, suppose that S is a maximal torus of G. Since each x_i commutes with S, it lies in $Z_G(S) = S$. □

2.2. The case of rank zero

There is a general finiteness result in this case:

PROPOSITION 2.2.1. *Let* K *be a compact group with finite center. (We do not assume that* K *is connected nor that the component of the identity is simply connected.) Up to conjugation, there are only finitely many almost commuting ordered* N*-tuples of rank zero in* K. *Thus, for each anti-symmetric* $N \times N$*-matrix* $C = (c_{ij})$ *with coefficients in* $\mathcal{C}K$, *the moduli space* $\mathcal{M}^0_K(C)$ *of conjugacy classes of almost commuting ordered* N*-tuples of rank zero of type* C *in* K *is a finite set.*

PROOF. The proof is by induction on N. The case $N = 1$ is deduced from the following lemma.

LEMMA 2.2.2. *Suppose that* $\mathcal{C}K$ *is finite. There are only finitely many conjugacy classes of elements* $g \in K$ *for which the center of* $Z(g)$ *is finite. In case* K *is connected and simply connected, and thus semi-simple, the center of* $Z_K(x)$ *is finite if and only if* x *is conjugate to the exponential of a vertex of an alcove* A.

PROOF. Fix a maximal torus $T \subseteq K^0$. Let $W(T, K)$ be the normalizer of T in K modulo its centralizer. It is a finite group, whose action on T is covered by a linear action on $\mathfrak{t}$.

Fix $g \in K$. According to [**17**], II §3 Proposition 2 (see also [**2**]) there is a regular element in $\mathfrak{g}$ fixed under $\mathrm{Ad}\, g$. Thus, after conjugation we can assume that g normalizes T and a positive Weyl chamber $C_0 \subseteq \mathfrak{t}$. This implies that $T_0 = (T^g)^0$ contains a regular point of T. The torus T_0 is a maximal torus of K^g and the normalizer of T_0 in K is contained in the normalizer of T. Let us suppose that the center of $Z(g)$ is finite. This implies that $W_0 = W(T_0, Z(g))$ acts on T_0 and that there are only finitely many points fixed by the action of W_0. Equivalently, W_0 acts on the quotient torus $\overline{T}_g = T/(\mathrm{Id} - \mathrm{Ad}\, g)T$ with only finitely many points fixed by W_0.

It suffices to show that there are only finitely many conjugacy classes of elements $g' \in K$ which (i) are congruent to g modulo K^0, (ii) such that $\mathrm{Ad}\, g'|T = \mathrm{Ad}\, g|T$ and (iii) have the property that $W_0' = W(T_0, K^{g'})$ is equal to W_0.

Any element $g' \in K$ satisfying (i) and (ii) is an element of the form tg for some $t \in T$. The conjugacy class of tg in K depends only on $[t] \in \overline{T}_g$.

For each $w \in W_0$ there is an element $h_w \in N_K(T_0) \subseteq N_K(T)$ such that $[h_w, g] = 1$ and $[h_w] = w \in W_0$. Suppose that $g' = tg$ satisfies (iii) for some $t \in T$. Then, for all $w \in W_0$ there exists $h'_w = t_w h_w$, $t_w \in T$, commuting with tg. Then

$$t_w h_w t g h_w^{-1} t_w^{-1} = tg$$

or

$$t_w {}^w(t) {}^g(t_w^{-1}) = t.$$

Thus $[t] \in \overline{T}_g$ is fixed by w for every $w \in W_0$. This means that there are only finitely many possibilities for $[t] \in T_g$ and hence only finitely many possibilities for the conjugacy class of tg in K.

In the case when K is connected and simply connected, $Z(x)$ is connected. Thus, it has a finite center if and only if it is semi-simple. This occurs exactly when x is conjugate to the exponential of a vertex of an alcove A. □

There is the following which we shall need later:

LEMMA 2.2.3. *Let K be connected and simply connected, let $(x_1, \ldots, x_N)$ be a rank zero subset of almost commuting elements of K, and suppose that $[x_1, x_i] = 1$ for all i. Then $Z(x_1)$ is semi-simple and x_1 is conjugate in K to the exponential of a vertex of an alcove.*

PROOF. The $(N-1)$-tuple $(x_2, \ldots, x_N)$ is a rank zero almost commuting $(N-1)$-tuple in $Z(x_1)$. Hence the center of $Z(x_1)$ is finite, so that $Z(x_1)$ is semi-simple. It follows from Lemma 2.2.2 that x_1 is conjugate to the exponential of a vertex of an alcove. □

PROOF OF PROPOSITION 2.2.1.
CASE $N = 1$. An almost commuting 1-tuple is a single element $g \in K$. Rank zero means that $Z(g)$ is finite and *a fortiori* that its center is finite. Thus, the result in this case is immediate from Lemma 2.2.2.
GENERAL CASE. We prove the general case by induction on N. Suppose that the result is known for all groups with finite center and for all almost commuting k-tuples for $k < N$. Consider an ordered almost commuting N-tuple $\mathbf{x} = (x_1, \ldots, x_N) \subseteq K$ of rank zero. Let $\hat{Z}(x_N)$ be the subgroup of elements of K whose commutator with x_N lies in $\mathcal{C}K$. Since $\mathbf{x} \subseteq \hat{Z}(x_N)$ and $\mathbf{x}$ is of rank zero, we see that the center of $\hat{Z}(x_N)$ must be finite. Let $\overline{K}$ be the quotient of K by its center and let $\overline{x}_N$ be the image of x_N in $\overline{K}$. Then we have an exact sequence

$$\{1\} \to \mathcal{C}K \to \hat{Z}(x_N) \to Z_{\overline{K}}(\overline{x}_N) \to \{1\}$$

and it follows that the center of $Z_{\overline{K}}(\overline{x}_N)$ is finite. Applying the previous lemma to $\overline{K}$, we see that there are only finitely many possibilities for $\overline{x}_N$ up to conjugation in $\overline{K}$. Hence, there are only finitely many possibilities for $x_N \in K$ up to conjugation. Let $\mathbf{x}' = (x_1, \ldots, x_{N-1})$. This is an ordered almost commuting $(N-1)$-tuple in $\hat{Z}(x_N)$. Consider the center of the centralizer Z of $\mathbf{x}'$ in $\hat{Z}(x_N)$. Clearly, $Z(\mathbf{x}) = Z \cap Z(x_N)$ is a subgroup of finite index of Z. Since the center of $Z(\mathbf{x})$ is finite, it follows that the center of Z is finite. This means that $\mathbf{x}'$ is of rank zero. Thus, by the inductive hypothesis, there are only finitely many possibilities for $\mathbf{x}'$ up to conjugation in $\hat{Z}(x_N)$, and hence only finitely many possibilities for $\mathbf{x}'$ in $\hat{Z}(x_N)$ up to conjugation by $Z(x_N)$. This completes the inductive step. □

2.3. The case of arbitrary rank

In this section we return to the group G, which is connected and simply connected and thus semi-simple. Let $C = (c_{ij})$ be an antisymmetric $N \times N$ matrix with coefficients in $\mathcal{C}G$.

Fix a subset $I \subseteq \Delta$. Suppose that L_I contains all the elements c_{ij}, $1 \leq i, j \leq N$. Let $\mathcal{M}^0_{L_I}(C) \subseteq \mathcal{M}_{L_I}(C)$ be the subset of conjugacy classes of rank zero almost commuting ordered N-tuples of C-type in L_I. Let $\widetilde{\mathcal{M}}^0_{L_I}(C)$ be a set of representatives for the finite set $\mathcal{M}^0_{L_I}(C)$.

Let $F_I = S_I \cap L_I$. This is a finite subgroup of the center of L_I. Consider the action of F_I^N on L_I^N given by

$$(f_1, \ldots, f_N) \cdot (y_1, \ldots, y_N) = (f_1 y_1, \ldots, f_N y_N).$$

Clearly, this operation does not change the pairwise commutators, nor does it change the centralizer in L_I of the subset. Hence, it defines an action of F_I^N on $\mathcal{M}^0_{L_I}(C)$.

Then we have a map

$$S_I^N \times \mathcal{M}^0_{L_I}(C)) \to \mathcal{M}_G(C)$$

which associates to $(s_1, \ldots, s_N) \times r$ the conjugacy class of the ordered N-tuple

$$\mathbf{x} = (s_1 \tilde{r}_1, \ldots, s_N \tilde{r}_N),$$

where $(\tilde{r}_1, \ldots, \tilde{r}_N)$ is the chosen representative in $\widetilde{\mathcal{M}}_G(C)^0$ for the element $r \in \mathcal{M}^0_{L_I}(C)$. It is clear from the definitions that this map is independent of the choice of representatives $\widetilde{\mathcal{M}}^0_{L_I}(C)$ for the conjugacy classes $\mathcal{M}^0_{L_I}(C)$ and that it factors to define a continuous map

$$p \colon S_I^N \times_{F_I^N} \mathcal{M}^0_{L_I}(C) \to \mathcal{M}_G(C).$$

The group $N_G(S_I)$ acts by conjugation normalizing S_I and hence L_I. Thus, it acts on $\mathcal{M}^0_{L_I}(C)$. The group $Z_G(S_I)$ acts on L_I by inner automorphisms of L_I and hence acts trivially on $\mathcal{M}^0_G(C)$. Thus, we have an induced action of $W(S_I, G) = N_G(S_I)/Z_G(S_I)$ on

$$S_I^N \times_{F_I^N} \mathcal{M}^0_{L_I}(C).$$

Clearly, the map p factors through this action.

THEOREM 2.3.1. *For $I \subseteq \Delta$, let $\mathcal{M}^I_G(C) \subseteq \mathcal{M}_G(C)$ be the subspace of conjugacy classes of ordered N-tuples $\mathbf{x}$ of C-type in G whose centralizer has a maximal torus which is conjugate to S_I. The map p induces a homeomorphism*

$$\overline{p} \colon \left(S_I^N \times_{F_I^N} \mathcal{M}^0_{L_I}(C))\right) / W(S_I, G) \to \mathcal{M}^I_G(C).$$

In particular, $\mathcal{M}^I_G(C)$ is a compact Hausdorff space and hence a closed subset of $\mathcal{M}_G(C)$.

COROLLARY 2.3.2. *For each $I \subseteq \Delta$ the subset $\mathcal{M}^I_G(C)$ is a union of components of $\mathcal{M}_G(C)$. In particular, the rank of $\mathbf{x}$ and the subset $I(\mathbf{x})$ are locally constant functions on $\mathcal{M}_G(C)$.*

PROOF OF COROLLARY 2.3.2. According to the theorem the subset $\mathcal{M}^I_G(C)$ is compact and hence is a closed subset of $\mathcal{M}_G(C)$. Since $\mathcal{M}_G(C)$ is a disjoint union of the $\mathcal{M}_G^I(C)$ for the various $I \subseteq \Delta$ and since these subsets are closed and finite in number, each is a union of components. □

PROOF OF THEOREM 2.3.1. Fix a subset $I \subseteq \Delta$. Suppose that $\mathbf{x}$ is an ordered N-tuple of C-type in G and that a maximal torus S for $Z(\mathbf{x})$ is conjugate to S_I. Conjugating $\mathbf{x}$ we can assume that $S = S_I$. Then $\mathbf{x} \subseteq Z(S_I) = S_I \cdot L_I$. Thus, we can write $\mathbf{x} = (s_1 y_1, \ldots, s_N y_N)$ with $s_i \in S_I$ and $y_i \in L_I$. It follows that $c_{ij} \in L_I$ for all $1 \leq i < j \leq N$ and that $\mathbf{x}' = (y_1, \ldots, y_N)$ is an ordered N-tuple of C-type in L_I. Since $S \cdot Z_{L_I}(\mathbf{x}') \subseteq Z(\mathbf{x})$, the N-tuple $\mathbf{x}'$ is of rank zero in L_I.

This shows that the map $\overline{p}$ is onto the subset $\mathcal{M}_G^I(C)$ of $\mathcal{M}_G(C)$. To prove that $\overline{p}$ is a homeomorphism onto $\mathcal{M}_G^I(C)$ we need only show that $\overline{p}$ is one-to-one. Suppose that $\overline{p}((s_1, \ldots, s_N), \mathbf{x}) = \overline{p}((s_1', \ldots, s_N'), \mathbf{x}')$ where $\mathbf{x} = (y_1, \ldots, y_N)$ and $\mathbf{x}' = (y_1', \ldots, y_N')$ are elements of $\widetilde{\mathcal{M}}_{L_I}^0(C)$. Then there is $g \in G$ such that

$$g(s_1 y_1, \ldots, s_N y_N) g^{-1} = (s_1' y_1', \ldots, s_N' y_N').$$

Thus $gS_I g^{-1}$ and S_I are both maximal tori for $Z = Z(s_1' y_1', \ldots, s_N' y_N')$ and hence there is an element in $h \in Z$ such that $hS_I h^{-1} = gS_I g^{-1}$. Replacing g by $h^{-1}g$ allows us to assume that $g \in N_G(S_I)$. Then $g\mathbf{x}g^{-1}$ is a rank-zero ordered N-tuple of C-type in L_I. Also, $gy_i g^{-1} = (gs_i^{-1} g^{-1} s_i') y_i'$, implying that $f_i = gs_i^{-1} g^{-1} s_i' \in F_I$. Let $f \in F_I^N$ be the given by $(f_1, \ldots, f_N)$. Then there is an element of L_I which conjugates $g\mathbf{x}g^{-1}$ to the given representative for the conjugacy class of $f \cdot \mathbf{x}'$ in $\mathcal{M}_{L_I}^0(C)$. This proves that $((s_1, \ldots, s_N), \mathbf{x})$ and $((s_1', \ldots, s_N'), \mathbf{x}')$ have the same image in $(S_I^N \times_{F_I^N} \mathcal{M}_{L_I}^0(C))/W(S_I, G)$. □

COROLLARY 2.3.3. *There is a component of $\mathcal{M}_G(C)$ consisting of conjugacy classes of ordered N-tuples $\mathbf{x}$ of C-type with $I(\mathbf{x}) = I$ if and only if* (i) *$c_{ij} \in L_I$ for all $1 \leq i, j \leq N$ and* (ii) *L_I contains an ordered N-tuple of C-type of rank zero. The set of these components is identified with $\left(\mathcal{M}_{L_I}^0(C)/F_I^N\right)/W(S_I, G)$. The dimension of each such component is $N \cdot \#(\Delta - I)$. For each component X of $\mathcal{M}_G^I(C)$ let $F_I^N(X) \subseteq F_I^N$ be the stabilizer of the corresponding point of $\mathcal{M}_{L_I}^0(C)$. Then X is homeomorphic to*

$$\left(S_I^N / F_I^N(X)\right) / W(S_I, G).$$

There is a similar but somewhat more involved statement of the corollary in case G is not connected.

CHAPTER 3

Some characterizations of groups of type A

3.1. Generalities on subroot systems

We begin with a couple of elementary, and well-known results about certain types of subroot systems. Let G be simple, and let $A \subseteq i\mathfrak{t}$ be the alcove containing the origin determined by the set of simple roots Δ. Recall that the walls of A correspond to the extended simple roots $\widetilde{\Delta}$. We then have the following well-known lemma, whose proof is left to the reader:

LEMMA 3.1.1. *Let $\tilde{x} \in A$ and let $x = \exp(2\pi i\tilde{x}) \in T$. Let $\Phi(x)$ be the set of $a \in \Phi$ such that $a(\tilde{x}) \in \mathbf{Z}$. Finally let $\widetilde{I}(x)$ be the set of extended simple roots a such that $\tilde{x}$ lies in the wall W_a of A corresponding to a. Then $\Phi(x)$ is a closed sub-root system of Φ, and $\widetilde{I}(x)$ is a set of simple roots for $\Phi(x)$. Thus $\widetilde{I}(x)$ corresponds to a proper subdiagram of $\widetilde{D}(G)$, and moreover every proper subdiagram of $\widetilde{D}(G)$ corresponds to a subset $\widetilde{I}(x)$ for some $x = \exp(2\pi i\tilde{x})$ with $\tilde{x} \in A$.*

There is the following lemma on the relationships of the coroot lattices.

LEMMA 3.1.2. *Let $\widetilde{I}$ be a proper subset of $\widetilde{\Delta}$, and let $Q^\vee_{\widetilde{I}}$ be the sublattice of $Q^\vee$ spanned by the coroots $a^\vee$, $a \in \widetilde{I}$. Let $k = \{\gcd g_a : a \notin \widetilde{I}\}$. Then the torsion subgroup of $Q^\vee/Q^\vee_{\widetilde{I}}$ is cyclic of order k, and a generator is*

$$\zeta = -\frac{1}{k}\sum_{a\in\widetilde{I}} g_a a^\vee.$$

COROLLARY 3.1.3. *Let $x = \exp(2\pi i\tilde{x})$ for some $\tilde{x} \in A$. Let $\widetilde{I}(x)$ be the subset of $\widetilde{\Delta}$ consisting of all $a \in \widetilde{\Delta}$ such that the corresponding wall W_a contains $\tilde{x}$. Then $\widetilde{I}(x)$ is a set of simple roots for $DZ(x)$. Moreover $\pi_1(DZ(x))$ is a cyclic group of order $k = \gcd_{a\in\widetilde{\Delta}-\widetilde{I}(x)}(g_a)$. A generator for the fundamental group, viewed as a central element in the simply connected covering group $\widetilde{DZ}(x)$, is $\exp(2\pi i\zeta)$ where*

$$\zeta = -\frac{1}{k}\sum_{a\in\widetilde{I}(x)} g_a a^\vee.$$

3.2. Action of $\mathcal{C}G$ on an alcove

Let Φ be an irreducible root system on V and let $A \subset V$ be the alcove determined by a set Δ of simple roots. The affine Weyl group $W_{\text{aff}}(\Phi)$ acts as a group of affine linear transformations of V with the alcove A as fundamental domain. For each element $c \in \mathcal{C}\Phi$ there is a unique point $\zeta_c \in A \cap P^\vee$ whose image in $\mathcal{C}\Phi = P^\vee/Q^\vee$ in c. In fact, ζ_c is a vertex of A and if $\{a = 0\}$ defines the wall opposite this vertex then, in Equation 1.2, the root integer h_a is equal to 1. This identifies $\mathcal{C}\Phi$ with the subset of $a \in \Delta$ for which $h_a = 1$.

Let $A' = A - \zeta_{c^{-1}}$. Then A' is another alcove containing 0. Hence there is a unique element $w_c \in W(\Phi)$ with the property that $w_c \cdot A = A'$. The map

$$\varphi_c(t) = w_c \cdot (t - \zeta_{c^{-1}}),$$

is an affine linear transformation of V carrying A to itself. We denote its fixed point set by V^c and we set $A^c = A \cap V^c$. The map $c \mapsto \varphi_c$ defines a homomorphism from $\mathcal{C}\Phi$ to the group of affine linear automorphisms of A. This map is called *the action of* $\mathcal{C}\Phi$ *on the alcove* A. The element $w_c \in W$ is called the *Weyl part of the action of* c. We define a homomorphism $\nu = \nu_A \colon \mathcal{C}\Phi \to W(\Phi)$ by associating to c the element w_c. Notice that ν is an injective homomorphism $\mathcal{C}\Phi \to W(\Phi)$. Its image is the stabilizer in $W(\Phi)$ of $\widetilde{\Delta}$ and acts simply transitively on the set of $a \in \widetilde{\Delta}$ such that $h_a = 1$. If A' is another alcove containing the origin, then $\nu_{A'}$ is conjugate to ν_A. More generally, for any alcove A', not necessarily containing the origin, there is an action of $\mathcal{C}(\Phi)$ on A' by elements of the affine Weyl group.

Notice that A^c contains the barycenter of the alcove A and hence is non-empty. Thus, the dimension of A^c is equal to the dimension of V^c which in turn is equal to the dimension of the fixed point subspace V^{w_c} for the linear action of w_c on V. The affine action φ_c permutes the codimension-one faces of A, and hence $w_c \in W$ permutes the roots in $\widetilde{\Delta}$ and induces a diagram automorphism of $\widetilde{D}(\Phi)$. In fact, every diagram automorphism of $\widetilde{D}(\Phi)$ is induced by a permutation of $\widetilde{\Delta}$ of the form $w_c \circ \sigma$, where $c \in \mathcal{C}\Phi$ and σ is an outer automorphism of Φ, which can be identified with a diagram automorphism of $D(\Phi) \subseteq \widetilde{D}(\Phi)$. Since there is only one integral linear relation among the $a \in \widetilde{\Delta}$ with positive coefficients with no common factor, namely $\sum_{a \in \widetilde{\Delta}} h_a a = 0$, the action of w_c preserves this relation. Thus, $h_a = h_{w_c \cdot a}$ for all $a \in \widetilde{\Delta}$. For the same reason, the action of w_c preserves the linear relation $\sum_{a \in \widetilde{\Delta}} g_a a^\vee$, i.e. $g_a = g_{w_c \cdot a}$ for all $a \in \widetilde{\Delta}$.

Now suppose that $\Phi = \Phi(T, G)$. Translating through the identifications we have an action $c \mapsto \varphi_c$ of $\mathcal{C}G$ on the alcove $A \subset i\mathfrak{t}$. For each $c \in \mathcal{C}G$ we denote by A^c the fixed point set of this action and we denote by $w_c \in W(T, G)$ the Weyl part of the action. Thus, we have a homomorphism $\nu_A \colon \mathcal{C}G \to W(T, G)$ given by $c \mapsto w_c$. Furthermore, if $\gamma_c \in N_G(T)$ projects to $w_c \in W$ and $x = \exp(2\pi i \tilde{x})$ for some $\tilde{x} \in A^c$, then

$$\gamma_c \cdot x \cdot \gamma_c^{-1} = c^{-1} x. \tag{3.1}$$

More generally, let H be a compact connected group with maximal torus T. Let $\Lambda = \pi_1(T)$. Then $(2\pi i)$ times the coroots inverse to the roots Φ_H of H with respect to T span a subspace $\mathfrak{d} \subseteq \mathfrak{t}$, and Φ_H is a root system on $i\mathfrak{d}$. Let $Q_H^\vee$ be the lattice spanned by the coroots and let $P_H^\vee$ be the corresponding coweight lattice. Then $2\pi i Q_H^\vee \subseteq \Lambda \subset \mathfrak{t}$ and $\pi_1(H) \cong \Lambda/2\pi i Q_H^\vee$. The preimage of T in the universal covering $\widetilde{H}$ of H is $\mathfrak{t}/2\pi i Q_H^\vee$. Lastly, if π is orthogonal projection from $\mathfrak{t}$ to $\mathfrak{d}$ under a Weyl invariant inner product, then $\pi(\Lambda) \subseteq 2\pi i P_H^\vee$. Thus there is an induced homomorphism

$$\Lambda/2\pi i Q_H^\vee \to P_H^\vee/Q_H^\vee.$$

We consider the decomposition $\mathcal{A}_H$ of $i\mathfrak{t}$ into alcoves under the walls of the affine Weyl group $W_{\text{aff}}(H)$. If $\mathfrak{z}$ is the Lie algebra of the center of H, then the alcoves in $\mathcal{A}_H$ are of the form $A' \times i\mathfrak{z}$ where A' is an alcove for the affine Weyl group of the derived group DH with respect to its maximal torus $T \cap DH$. Thus, the alcoves in $\mathfrak{i}\mathfrak{t}$ are compact if and only if H is semi-simple. Since $W_{\text{aff}}(H) = W_{\text{aff}}(DH)$, it follows that this group acts simply transitively on the set of alcoves in $\mathfrak{i}\mathfrak{t}$.

Fix a set of simple roots Δ_H for the root system of H with respect to T and let $\widetilde{\Delta}_H$ be the corresponding set of extended roots. Let $A \subseteq \mathfrak{i}\mathfrak{t}$ be the alcove determined by Δ_H. This alcove contains the origin. Conversely, given any alcove A' containing the origin, it corresponds to a set of simple roots Δ' for Φ_H. Exactly as in the semi-simple case, the center of $\widetilde{H}$ acts as a group of affine linear isometries of A. The linear part of this automorphism defines a homomorphism $\nu\colon \mathcal{C}\widetilde{H} \to W(H)$. The only difference with the semi-simple case is that ν is not injective – its kernel is the identity component of $\mathcal{C}\widetilde{H}$. For each $c \in \mathcal{C}\widetilde{H}$, we let $w_c = \nu(c)$. The element w_c acts on the roots of H, normalizing the set $\widetilde{\Delta}_H$. The action on $\widetilde{\Delta}_H$ is the one induced by the action of c on the walls of A. Using the homomorphism $\Lambda \to \mathcal{C}\widetilde{H}$ described above, we can also view Λ as acting on A.

LEMMA 3.2.1. *Let H be a compact group with maximal torus T and let $\varphi\colon \mathfrak{i}\mathfrak{t} \to \mathfrak{i}\mathfrak{t}$ be an affine linear map whose translation part is given by an element $v \in \mathfrak{i}\mathfrak{t}$ with $\exp(2\pi i v) = c \in \mathcal{C}\widetilde{H}$ and whose linear part is an element of the Weyl group $W(T,H)$. If there is an alcove $A \subseteq \mathfrak{i}\mathfrak{t}$ for the affine Weyl group of H stabilized by φ, then φ is the action of c on this alcove. In particular, its linear part is w_c.*

PROOF. Consider the composition of φ^{-1} and the affine linear map which is the action of c on A. The translation part of this map is given by an element of $v' \in \mathfrak{i}\mathfrak{t}$ with $\exp(2\pi i v') = 1 \in \widetilde{H}$. Hence, v' is contained in the coroot lattice for $\widetilde{H}$. The linear part is a composition of elements of the Weyl group and hence is an element of the Weyl group. That is to say this composition is an element of the affine Weyl group of H. Of course, it sends A to A. This means that it is the trivial element of the affine Weyl group. □

For future reference, we shall also need the following lemma on the stabilizer of a point of T under the action of the Weyl group.

LEMMA 3.2.2. *Let H be a compact group with $T \subset H$ a maximal torus. Fix an alcove $A \subset \mathfrak{i}\mathfrak{t}$. Let $t \in T$ and let $\tilde{t} \in A$ satisfy $\exp(2\pi i\tilde{t}) = t$. Let $W = W(T,H)$ and let $W(t)$ be the Weyl group generated by the roots a such that $a(\tilde{t}) \in \mathbf{Z}$, in other words by the root system $\Phi_H(t)$. Then the map ν induces an isomorphism*

$$\mathrm{Stab}_{\Lambda/2\pi i Q_H^\vee}(\tilde{t}) \cong \mathrm{Stab}_W(t)/W(t).$$

Moreover, if I is the image of $\mathrm{Stab}_{\Lambda/2\pi i Q_H^\vee}(\tilde{t})$ under ν, then $I = \mathrm{Stab}_W(t) \cap \mathrm{Im}\,\nu$ and $\mathrm{Stab}_W(t) \cong W(t) \rtimes I$.

PROOF. Suppose $c \in \Lambda/2\pi i Q_H^\vee$. If φ_c stabilizes $\tilde{t}$, then $\tilde{t} = w_c(\tilde{t} - \zeta_{c^{-1}})$ for some $\zeta_{c^{-1}}$ such that $\exp(2\pi i \zeta_{c^{-1}}) = c^{-1}$, and so $w_c(t) = t$. Thus ν defines a homomorphism from $\mathrm{Stab}_{\Lambda/2\pi i Q_H^\vee}(\tilde{t})$ to $\mathrm{Stab}_W(t)$. If in the above notation $w_c \in W(t)$, then by Lemma 3.1.1, w_c is in the group generated by reflections in the roots which are integral on $\tilde{t}$. Thus $w_c(\tilde{t}) = \tilde{t} + \lambda$, where $\lambda \in Q_H^\vee$. From

$$w_c(\tilde{t}) = \tilde{t} + w_c(\zeta_{c^{-1}}) = \tilde{t} + \lambda,$$

it follows that $w_c(\zeta_{c^{-1}}) \in Q_H^\vee$ and hence that $\zeta_{c^{-1}} \in Q_H^\vee$. Thus, $c^{-1} \in \Lambda/2\pi i Q_H^\vee$ is trivial, so that c is trivial as well. This shows that $\nu|\mathrm{Stab}_{\Lambda/2\pi i Q_H^\vee}(\tilde{t})$ is injective, and, if I is its image, then $I \cap W(t) = \{1\}$.

Now suppose that $w \in W$ fixes t. Thus $w(\tilde{t}) = \tilde{t} + \tilde{c}$ for some $\tilde{c}$ such that $2\pi i\tilde{c} \in \Lambda$. Hence A and $A' = w(A) - \tilde{c}$ are two alcoves, both of which contain $\tilde{t}$. After transforming A' by an element in the group generated by reflections in the walls of A containing $\tilde{t}$, we can then assume that $A' = A$. By Lemma 3.2.1, it follows that $w = w_c$. This says that every element of $\mathrm{Stab}_W(t)$ can be written as a product of an element in $\mathrm{Im}\,\nu$ times an element of the group generated by reflections in the walls of A containing $\tilde{t}$. But by Lemma 3.1.1, this second group is exactly $W(t)$. This proves that $I = \mathrm{Stab}_W(t) \cap \mathrm{Im}\,\nu$ and that $\mathrm{Stab}_W(t) = W(t) \cdot I$. Since $W(t)$ is a normal subgroup of $\mathrm{Stab}_W(t)$ and the product decomposition is unique, by the first paragraph of the proof, we see that $\mathrm{Stab}_W(t) \cong W(t) \rtimes I$. □

Clearly, if $\lambda \in \Lambda/2\pi i Q_H^\vee$ has infinite order, then it does not stabilize any point. Thus $\mathrm{Stab}_{\Lambda/2\pi i Q_H^\vee}(\tilde{t})$ is identified with the stabilizer of $\tilde{t}$ in the torsion subgroup of $\Lambda/2\pi i Q_H^\vee$.

Finally, we note that there are analogues of the the previous two lemmas, with essentially identical proofs, in the case of a vector space $V = U \oplus Z$, a root system Φ, not necessarily reduced, on U, and a lattice $\Lambda \subseteq V$ such that

$$Q^\vee(\Phi) \subseteq \Lambda \cap U \subseteq \pi_U(\Lambda) \subseteq P^\vee(\Phi),$$

where π_U is projection to the factor U.

3.3. A first characterization of groups of type A

LEMMA 3.3.1. *Let Φ be irreducible, but not necessarily reduced, with d as highest root and let $e = \sum_{a\in\Delta} e_a a$ be the highest short root. (Of course, $d = e$ if and only if Φ is simply laced.)*

(1) *d and e are orthogonal to all but either one or two simple roots.*
(2) *The following are equivalent:*
  (i) *R is of type A_n for some $n \geq 2$;*
  (ii) *d is not orthogonal to two simple roots;*
  (iii) *e is not orthogonal to two simple roots.*
(3) *If R is not of type A_n for any $n \geq 1$, then $e = \varpi_a$ for some $a \in \Delta$.*

PROOF. We have

$$2 = n(d,d) = \sum_{a\in\Delta} h_a n(a,d) = n(e,e) = \sum_{a\in\Delta} e_a n(a,e).$$

The coefficients h_a and e_a, are positive integers. Since d is the highest root, resp. e is the highest short root, the $n(a,d)$, resp. $n(a,e)$ are also nonnegative. Statement 1 follows.

Clearly, (i) implies (ii) and (iii). Moreover, since the passage to an inverse system is conformal and permutes the roles of d and e, (ii) is equivalent to (iii). It remains to see that (ii) implies (i). Assume (ii) holds. Let a, b be the simple roots not orthogonal to d. There exists a root f which is a sum of distinct simple roots including a, b. (Namely, f is the sum of the simple roots corresponding to the nodes of the interval connecting a and b in the Dynkin diagram for Φ.) Then $n(f,d) = 2$. But since d is a root at least as long as all other roots, the Cartan integer $n(f,d) = 2$ implies that $f = d$, and (i) follows. This proves Part (2).

By Statement 2, there is a unique simple root a for which $n(e,a) \neq 0$ meaning that e is a multiple of ϖ_a. Of course, $n(e,a) \geq 0$. Since e is short, $n(e,a) \in \{1,2\}$, with $n(e, a) = 2$ if and only if $a = e$. Since R is not of type A_1, this can never happen, proving Part (3). $\square$

PROPOSITION 3.3.2. (1) *Let G be simple. Then there exists a fundamental weight ϖ_a such that $\varpi_a(\mathcal{C}G) = 1$ if and only if G is not of type A.*

(2) *Let $G = \prod_{i=1}^{s} G_i$ be the decomposition of G into simple factors, and let $c \in \mathcal{C}G$ be the product $c = c_1 \cdots c_s$ with $c_i \in \mathcal{C}G_i$. Let $c = \exp(2\pi i \sum_{a\in\Delta} \lambda_a a^\vee)$ for $\lambda_a \in \mathbf{R}$. If no λ_a is integral, then for each $i, 1 \le i \le s$, we have $G_i = SU(n_i)$ for some $n_i \ge 2$ and the element c_i generates $\mathcal{C}G_i$.*

(3) *With $c \in \mathcal{C}G$ of the form $\exp(2\pi i \sum_{a\in\Delta} \lambda_a a^\vee)$ as above, suppose that G is of type A and that c generates $\mathcal{C}G$. Then*

$$\{\lambda_a \bmod \mathbf{Z}, a \in \Delta\} = \left\{\frac{k}{n+1}, 1 \le k \le n\right\}.$$

In particular, no λ_a is integral.

PROOF. (1) If G is simple and not of type A, then by Lemma 3.3.1 there is a fundamental weight ϖ_a which is a root and hence kills $\mathcal{C}G$. Conversely, if $G = SU(n+1)$, the fundamental representations are the exterior powers $\bigwedge^i \mathbf{C}^{n+1}$, $1 \le i \le n$, of the standard representation, and their highest weights are nontrivial on $\mathcal{C}SU(n+1)$.

If $c = \exp(2\pi i \sum_a \lambda_a a^\vee)$, then $\varpi_a(c) = \exp(2\pi i \lambda_a)$. Thus, (2) follows from (1) and the fact, again easily verified by direct inspection, that, for a proper subgroup C of the center of $SU(n)$, there is a fundamental weight which kills C.

Finally, (3) follows by examining a generator for the center of $SU(n+1)$. $\square$

The proof actually establishes the following:

ADDENDUM 3.3.3. Let Φ be a not necessarily reduced root system with irreducible factors Φ_i. Let Δ be a set of simple roots for Φ. Suppose that $\zeta = \sum_{a\in\Delta} \lambda_a a^\vee \in P^\vee(\Phi)$, where $\lambda_a \in \mathbf{Q}$. Then no λ_a is integral if and only if every Φ_i is of type A, and, for every i, the projection of $\zeta \in P^\vee(\Phi)/Q^\vee(\Phi) \cong \bigoplus_i P^\vee(\Phi_i)/Q^\vee(\Phi_i)$ to the factor $P^\vee(\Phi_i)/Q^\vee(\Phi_i)$ generates this factor.

3.4. Subgroups associated with elements of the center

Let $c \in \mathcal{C}G$ and let $\lambda \in i\mathfrak{t}$ be such that $\exp(2\pi i\lambda) = c$. Write

$$\lambda = \sum_{a\in\Delta} \lambda_a a^\vee. \tag{3.2}$$

Then the λ_α are rational numbers. Let $\Delta(c) = \{a \in \Delta | \lambda_a \notin \mathbf{Z}\}$. The set $\Delta(c)$ depends only on c and not on the choice of a lift λ. In the notation of Lemma 2.1.1, we set $\mathfrak{t}_c = \mathfrak{t}_{\Delta(c)}$, $S_c \subset T$ equal to the subtorus with $\mathrm{Lie}(S_c) = \mathfrak{t}_c$, and $L_c = L_{\Delta(c)} = DZ(S_c)$.

LEMMA 3.4.1. *With the previous notation we have*

(1) *$\varpi_a(c) = 1$ if and only if $a \notin \Delta(c)$.*

(2) *$c \in L_c$.*

(3) *If $I \subseteq \Delta$ has the property that $c \in L_I$, then $\Delta(c) \subseteq I$, and hence $L_c \subseteq L_I$.*

PROOF. Recall that $\varpi_a(c) = \exp(2\pi i\lambda_a)$ and that $2\pi i a^\vee \in \mathfrak{t} \cap \mathrm{Lie}\, L_I$ if and only if $a \in \Delta(c)$. From these facts (1) and (2) are clear.

If $c \in L_I$, then $c = \exp(2\pi i\lambda')$ for some element λ' in the real linear span of the coroots $a^\vee$ for $a \in \Delta_I$. Since the element λ' differs by an element of the coroot lattice from λ, we see that $\lambda_a \in \mathbf{Z}$ for all $a \notin I$. That is to say $\Delta(c) \subseteq I$. $\square$

3.5. A further characterization of products of groups of type A

PROPOSITION 3.5.1. *Let $G = \prod_{i=1}^s G_i$, with the G_i simple. Let $c \in \mathcal{C}G$. The following conditions are equivalent.*

(1) *The fixed point set T^{w_c} of the w_c-action on T is finite.*
(2) *For each i, $1 \leq i \leq s$, the group $G_i = SU(n_i)$ and the projection c_i of c into G_i generates $\mathcal{C}G_i$.*
(3) *$\Delta(c) = \Delta$, i.e. in Equation 3.2, no coefficient λ_a is integral.*

PROOF. We have $c = c_1 \cdots c_s$ with $c_i \in \mathcal{C}G_i$, $w_c = w_{c_1} \cdots w_{c_s}$, and $A = A_1 \times \cdots \times A_s$ where A_i is an alcove for G_i. The condition that T^{w_c} be finite implies that for each i the action of c_i on the alcove A_i permutes transitively the vertices of A_i. Hence, every vertex of the alcove A_i is an element of the center of G_i. This means that the highest root is the sum of the simple roots. The only groups with this property are groups of type A. This shows that (1) implies (2).

It is clear that (2) implies (1) and (3). The fact that (3) implies (2) follows from Proposition 3.3.2 and Lemma 3.4.1. $\square$

COROLLARY 3.5.2. *Let $c \in \mathcal{C}G$, and let L be a subgroup of G of the form L_I. Then $L = L_c$ if and only if*

(1) *$c \in L$;*
(2) *L is a product of simple factors $L_i \cong SU(n_i)$ for some n_i;*
(3) *The projection of c to L_i generates the center of L_i.*

PROOF. First suppose that $L = L_c$. By Section 3.4, $\Delta(c)$ is a set of simple roots for L_c and $c \in L_c$. Of course, in the expression $c = \exp(2\pi i \sum_{a\in\Delta(c)} \lambda_a a^\vee)$ all the coefficients λ_a are non-integral. Hence by Proposition 3.5.1 L_c is a product of groups $\prod_{i=1}^s L_i$ where L_i is isomorphic to $SU(n_i)$ for some integer $n_i \geq 2$ and $c = c_1 \cdots c_s$ where c_i generates the center of L_i.

Conversely, suppose that $L = L_I$ for some I, that $c \in L$ and that L is a product of groups $\prod_{i=1}^s L_i$ where L_i is isomorphic to $SU(n_i)$ for some integer $n_i \geq 2$ and $c = c_1 \cdots c_s$ where c_i generates the center of L_i. By Lemma 3.4.1, $\Delta(c) \subseteq I$. On the other hand, by Part (3) of Proposition 3.3.2, when we write $c = \exp(2\pi i \sum_a \lambda_a a^\vee)$ no coefficient λ_a for $a \in I$ is integral, and hence $I \subseteq \Delta(c)$. Thus $I = \Delta(c)$ and so $L = L_c$. $\square$

DEFINITION 3.5.3. Fix an element $c \in \mathcal{C}G$. Let w_c be the Weyl part of the action of c on the alcove A. Let T^{w_c} be the fixed points of the action of w_c on T. Let S^{w_c} be the identity component of T^{w_c}, and let $\mathfrak{t}^{w_c}$ be the fixed subspace for the action of w_c on $\mathfrak{t}$. Clearly, $\mathrm{Lie}(S^{w_c}) = \mathfrak{t}^{w_c}$.

PROPOSITION 3.5.4. *The torus S^{w_c} is conjugate to S_c.*

PROOF. Let $\mathfrak{t}_{L_c} = \mathfrak{t} \cap \mathrm{Lie}(L_c)$. It is the Lie algebra of a maximal torus for L_c. Let A' be the alcove in $i\mathfrak{t}_{L_c}$ associated with the set of simple roots $I_c \subseteq \Delta$ for L_c with respect to $\mathfrak{t}_{L_c}$. Let φ'_c be the action of c on A'. By Proposition 3.5.1 φ'_c fixes a unique point, say $\hat{p}$, of A', and $\hat{p}$ is the product of the barycenters of the factors of A'. Let $\widetilde{\varphi}_c$ be the extension of φ'_c to $i\mathfrak{t} = i\mathfrak{t}_{L_c} \oplus i\mathfrak{t}_c$ by the identity on $i\mathfrak{t}_c$.

A root of G which is integral on the affine space $i\mathfrak{t}_c + \hat{p}$ must vanish on $i\mathfrak{t}_c$ and hence be a root of L_c. But since lies in the interior of an alcove, it follows that there are no roots of G taking integral values on $i\mathfrak{t}_c + \hat{p}$. Thus there is an open dense subset of $\tilde{x} \in i\mathfrak{t}_c + \hat{p}$ such that $\exp(2\pi i\tilde{x})$ is a regular element of G. In particular, there is $v \in i\mathfrak{t}$ with the following three properties:

(1) $\exp(2\pi iv)$ is a regular element for G;
(2) v is fixed by $\widetilde{\varphi}_c$;
(3) the unique alcove $A \subset i\mathfrak{t}$ for the affine Weyl group of $\Phi(T, G)$ containing v also contains $\hat{p}$.

The point $\hat{p}$ lies in the alcove $A' \subset i\mathfrak{t}_{L_c}$, and A' contains the origin. Condition 3 above implies that A contains the origin. It follows from Conditions 1 and 2 that $\widetilde{\varphi}_c$ sends A to itself. By Lemma 3.2.1 we see that $\widetilde{\varphi}_c$ is the action of c on A. Thus, the fixed point set of the Weyl part w_c' of φ_c is exactly $\mathfrak{t}_c$ and exponentiates onto S_c. The proposition now follows since the Weyl part w_c of the action of c on A is conjugate to w_c', and hence S^{w_c} is conjugate to S_c. □

REMARK 3.5.5. There is an analogue of this result for an abstract root system Φ on V. Let Δ be a set of simple roots for Φ and let $A \subset V$ be the alcove determined by Δ. Let $c \in \mathcal{C}\Phi$ and let $\tilde{c} \in P^\vee$ be a lifting of c. We write

$$\tilde{c} = \sum_{a \in \Delta} \lambda_a a^\vee$$

and set $V_c = \cap_{a|\lambda_a \notin \mathbf{Z}} \mathrm{Ker}(a)$. Then V_c and V^{w_c} are conjugate under $W(\Phi)$.

3.6. A consequence of Proposition 3.5.1

THEOREM 3.6.1. *With notation as above, fix an integer $k > 1$ dividing at least one of the g_a for $a \in \widetilde{\Delta}$. Let $\widetilde{I}(k) = \{a \in \widetilde{\Delta} : k \nmid g_a\}$. Let $H(k)$ be the closed, connected subgroup of G whose complexified Lie algebra is generated by $\{(\mathfrak{g}^a \oplus \mathfrak{g}^{-a})\}_{a \in \widetilde{I}(k)}$. Then $H(k)$ is isomorphic to*

$$\left(\prod_{i=1}^r H_i\right) / \langle c \rangle$$

where:

(1) *c has order k;*
(2) *for each i, $1 \leq i \leq r$, the group H_i is isomorphic to $SU(n_i)$ for some $n_i | k$;*
(3) *for at least one i between 1 and r we have $n_i = k$;*
(4) *$c = \prod_{i=1}^r c_i$ and $c_i \in H_i$ generates the center of H_i.*

PROOF. The group $H(k)$ is a semi-simple subgroup of G for which $\widetilde{I}(k)$ is a set of simple roots. Let $\mathfrak{t}_{H(k)} \subseteq \mathfrak{t}$ be the subspace spanned by $2\pi i a^\vee$ for $a \in \widetilde{I}(k)$. Then $\mathfrak{t}_{H(k)}$ is the Lie algebra of a maximal torus of $H(k)$. The element

$$\lambda = \frac{1}{k} \sum_{a \in \widetilde{I}(k)} g_a a^\vee \tag{3.3}$$

is contained in $it_{H(k)}$ and is also in $Q^\vee$ since it is equal to

$$-\frac{1}{k}\sum_{a\in\widetilde{\Delta}-\widetilde{I}(k)} g_a a^\vee$$

and by definition $k|g_a$ for every $a\in\widetilde{\Delta}-\widetilde{I}(k)$. This means that every root of $H(k)$ takes integral values on λ, and hence that $2\pi i\lambda$ exponentiates to an element c contained in the center of the simply connected form $\widetilde{H}(k)$ of $H(k)$. By Equation 3.3, $c^k=1$.

The definition of λ implies that, when λ is expressed as a linear combination of the basis $a^\vee, a\in\widetilde{I}(k)$, all the coefficients of the simple coroots for $H(k)$ are non-integral. Hence, by Proposition 3.3.2, $\widetilde{H}(k)$ is a product $\prod_{i=1}^r H_i$, where for each i the group H_i is isomorphic to $SU(n_i)$ for some $n_i\geq 2$, and c is of the form $c_1\cdots c_r$ where for each i the element c_i generates the center of H_i. Since $c^k=1$, each of the c_i has order dividing k. Since c_i generates the center of H_i, its order is n_i. We conclude that $n_i|k$ for each i.

Consider now the component of $\widetilde{D}(k)$ that contains $\tilde{a}=-\tilde{d}$. (Since $g_{\tilde{a}}=1$, we have $\tilde{a}\in\widetilde{D}(k)$.) We index the H_i so that this component corresponds to H_1. Since the expression for λ as a linear combination of coroots has $1/k$ as the coefficient of $\tilde{a}$, it follows that the order of c_1 in H_1 is divisible by k, i.e., that $k|n_1$. Since we have already shown the opposite divisibility, it must be the case that $n_1=k$, showing that H_1 is isomorphic to $SU(k)$. Moreover, the order of c is divisible by k and hence is equal to k, and k is the least common multiple of the n_i.

The fundamental group of $H(k)$ is cyclic, by Lemma 3.1.2, and contains the element c, which is of order k. Thus k divides the order of $\pi_1(H(k))$. On the other hand, $\pi_1(H(k))$ is identified with a cyclic subgroup of $\prod_i \mathcal{C}SU(n_i)=\prod_i \mathbf{Z}/n_i\mathbf{Z}$, and hence its order divides the least common multiple of the n_i, namely k. Since c has order k, it generates the fundamental group of $H(k)$. □

COROLLARY 3.6.2. *If N is the maximal value for g_a for $a\in\Delta$, then for each k, $1\leq k\leq N$, there is at least one $a\in\widetilde{\Delta}$ for which $g_a=k$. In fact there is a simply laced chain of length N in $\widetilde{D}(G)$ containing $\tilde{a}$ as one end so that the g_a, in order, along this chain are $1,2,\ldots,N$.*

COROLLARY 3.6.3. *If Φ is an irreducible root system and k is a positive integer dividing at least one of the coroot integers g_a, then $\gcd\{\, g_a : k|g_a \,\}=k$.*

3.7. Application to generalized Cartan matrices and affine diagrams

In this section, we apply the above results on groups of type A to establish numerology concerning coroot integers, root integers, and more generally integers which are the coefficients of linear relations for nodes of diagrams of affine type. This numerology will be crucial for the proof of the Clockwise Symmetry result.

Let Ψ be an irreducible, but not necessarily reduced, root system and suppose that Υ is a set of simple roots for Ψ. As in the case of reduced root systems, there is the extended set of simple roots $\widetilde{\Upsilon}$, obtained by adding minus the highest root to Υ. There is an extended Dynkin diagram $\widetilde{D}(\Psi)$ whose nodes are the extended set of simple roots $\widetilde{\Upsilon}$, and whose bonds, with their multiplicities and arrows are determined by the Cartan integers $n(a,b)$ for $a,b\in\widetilde{\Upsilon}$ exactly as in the case of the ordinary diagram. Dually, there is the extended diagram $\widetilde{D}^\vee(\Psi)$ associated with the coroots inverse to the extended roots; it is obtained from $\widetilde{D}(\Psi)$ by reversing the direction of every arrow.

More generally, suppose that we are given a finite set $\widetilde{\Upsilon}$ and an integral matrix $N = (n(a,b))$, where $a, b \in \widetilde{\Upsilon}$. The matrix N is called a *generalized Cartan matrix* if:

(1) $n(a,a) = 2$ for all $a \in S$.
(2) $n(a,b) \leq 0$ for all $a \neq b \in S$.
(3) $n(a,b) = 0$ implies $n(b,a) = 0$.

If in addition the set $\widetilde{\Upsilon}$ cannot be divided into two disjoint non-empty subsets S_1 and S_2 such that $n(a,b) = 0$ for all $(a,b) \in S_1 \times S_2$, then we call N *indecomposable*. We will assume throughout that N is indecomposable.

We can form a diagram associated with a generalized Cartan matrix whose nodes are indexed by $\widetilde{\Upsilon}$. It has bonds with multiplicities and arrows determined by the same rules as in the case of Dynkin diagrams. Because of indecomposability, the diagram associated with a generalized Cartan matrix is connected. Notice that one can reconstruct the generalized Cartan matrix from its diagram.

Given a real vector space $\widetilde{V}$ of dimension $d+1$ and a positive semidefinite bilinear form $\langle \cdot, \cdot \rangle$, suppose that $\widetilde{\Upsilon}$ is a basis of $\widetilde{V}$ such that $\langle v, v \rangle \neq 0$ for all $v \in \widetilde{\Upsilon}$. Then we can define $n(v,w)$ by the usual formula

$$n(v,w) = \frac{2\langle v,w \rangle}{\langle w,w \rangle}. \tag{3.4}$$

If $N = (n(v,w))$ is a generalized Cartan matrix and there is a vector $u = \sum_{v \in \widetilde{\Upsilon}} n_v v$ in $\widetilde{V}$ such that $n_v > 0$ for all v and such that $\langle u, x \rangle = 0$ for all $x \in \widetilde{V}$, then the generalized Cartan matrix is said to be *of affine type*. In this case the associated diagram is called *an affine diagram*, and, according to a theorem of Kac [**10**], the diagram is either the extended root or the extended coroot diagram of a possibly non-reduced root system, and the coefficients of the vector u are a fixed positive integral multiple of the (root or coroot) integers on the extended diagram. Moreover, every proper subdiagram is the Dynkin diagram of a root system. Let $V(\widetilde{\Upsilon})$ be the quotient of $\widetilde{V}$ by the one-dimensional radical of the semidefinite form and let $Q(\widetilde{\Upsilon})$ be the lattice in $V(\widetilde{\Upsilon})$ spanned by the image of $\widetilde{\Upsilon}$. Note that $V(\widetilde{\Upsilon})$ is a d-dimensional vector space with a positive definite inner product. Projection induces a bijection from $\widetilde{\Upsilon}$ to a spanning set of $V(\widetilde{\Upsilon})$ of cardinality d, and the Cartan integers are determined by inner products of their images in $V(\widetilde{\Upsilon})$ by Equation 3.4. It follows from the theorem of Kac that the reflections in the $v \in \widetilde{\Upsilon}$ generate a Weyl group acting by isometries on $V(\widetilde{\Upsilon})$ and that the lattice $Q(\widetilde{\Upsilon})$ spanned by $\widetilde{\Upsilon}$ is invariant under this group. It also follows that there is one linear relation between the vectors of $\widetilde{\Upsilon}$ and that the coefficients of this relation can be chosen to be positive integers.

This construction can be reversed: suppose that V is a vector space of dimension d with a positive definite inner product and let $\widetilde{\Upsilon}$ be a subset of cardinality $d+1$ spanning V such that the Cartan numbers defined by Equation 1.1 are integers and determine an indecomposable generalized Cartan matrix. Then this matrix is of affine type.

For example, if G is simple of rank r, then $\widetilde{\Delta}^\vee$ is a subset of $i\mathfrak{t}$ of cardinality $r+1$. This embedding induces an identification of $V(\widetilde{\Delta}^\vee)$ with $i\mathfrak{t}$, and further identifies the coroot lattice $Q^\vee$ with the lattice $Q(\widetilde{\Delta}^\vee)$ and the Weyl group of G with the group generated by the reflections in $\widetilde{\Delta}^\vee$.

In this general context of affine diagrams, we have the following generalization of Corollary 3.6.2.

PROPOSITION 3.7.1. *Let $D(\widetilde{\Upsilon})$ be a connected affine diagram whose nodes are indexed by $v \in \widetilde{\Upsilon}$. Let $\mathbf{n}\colon \widetilde{\Upsilon} \to \mathbf{N}$ be a function such that $\sum \mathbf{n}(v)v = 0$ in $V(\widetilde{\Upsilon})$. We denote $\mathbf{n}(v)$ by n_v. Let k be a positive integer which divides at least one of the n_v. Let $\widetilde{I}(\mathbf{n},k) = \{\, v : k \nmid n_v \,\} \subseteq \widetilde{\Upsilon}$. Then there exist cyclic subgroups $C_i \subseteq \mathbf{Z}/k\mathbf{Z}$, not necessarily distinct, and a bijection*

$$\phi_k\colon \widetilde{I}(\mathbf{n},k) \to \coprod_i (C_i - \{0\})$$

such that $n_v \equiv \phi_k(v) \bmod k$ for all $v \in \widetilde{I}(\mathbf{n},k)$.

PROOF. Let

$$\zeta = -\frac{1}{k} \sum_{v\in \widetilde{I}(\mathbf{n},k)} n_v v.$$

Then ζ lies in the $\mathbf{R}$-span $V(\widetilde{I}(\mathbf{n},k))$ of $\widetilde{I}(\mathbf{n},k)$ in $V(\widetilde{\Upsilon})$ as well as in $Q(\widetilde{\Upsilon})$. Since $\widetilde{I}(\mathbf{n},k)$ is a proper subset of $\widetilde{\Upsilon}$, it is a set of simple roots for a root system on $V(\widetilde{I}(\mathbf{n},k))^*$ with the property that the given inner product on $V(\widetilde{\Upsilon})$ restricts to a Weyl invariant inner product on $V(\widetilde{I}(\mathbf{n},k))$. Let Ψ be the inverse root system on $V(\widetilde{I}(\mathbf{n},k))$, so that $\widetilde{I}(\mathbf{n},k)$ is a set of simple coroots for Ψ. Since $\zeta \in Q(\widetilde{\Upsilon})$, ζ has integral inner product with the lattice spanned by $\widetilde{I}(\mathbf{n},k)$ and hence $\zeta \in P^\vee(\Psi)$. Moreover, all of the coefficients of ζ are non-integral with respect to the set of simple coroots given by $\widetilde{I}(\mathbf{n},k)$. Thus, by Addendum 3.3.3, Ψ is a product of irreducible root systems Ψ_i of type A_{N_i} for some integers N_i and ζ projects into each factor $P^\vee(\Psi_i)/Q^\vee(\Psi_i)$ as a generator. Since $k\zeta \in Q^\vee(\Psi)$, it follows that $(N_i+1)|k$. Let C_i be the cyclic subgroup of $\mathbf{Z}/k\mathbf{Z}$ of order N_i+1. The result now follows immediately from Part 3 of Proposition 3.3.2. □

3.8. Numerology of clockwise symmetry

Let $\widetilde{\Upsilon}$ be a finite set and let $\mathbf{n}\colon \widetilde{\Upsilon} \to \mathbf{N}$ be a function. We denote $\mathbf{n}(v)$ by n_v. For every positive integer k which divides at least one of the n_v, let $\widetilde{I}(\mathbf{n},k) = \{\, v : k \nmid n_v \,\} \subseteq \widetilde{\Upsilon}$. We suppose throughout that the pair $(\widetilde{\Upsilon}, \mathbf{n})$ satisfies the conclusions of Proposition 3.7.1:

ASSUMPTION 3.8.1. *For every positive integer k which divides at least one of the n_v, there exist cyclic subgroups $C_i \subseteq \mathbf{Z}/k\mathbf{Z}$, not necessarily distinct, and a bijection*

$$\phi_k\colon \widetilde{I}(\mathbf{n},k) \to \coprod_i (C_i - \{0\})$$

such that $n_v \equiv \phi_k(v) \bmod k$.

Let $n_0 = \gcd\{n_v : v \in \widetilde{\Upsilon}\}$. We call the pair $(\widetilde{\Upsilon}, \mathbf{n})$ *reduced* if $n_0 = 1$. In general, if we define $n'_v = n_v/n_0$, the pair $(\widetilde{\Upsilon}, n')$ also satisfies the conclusions of Assumption 3.8.1 and is reduced.

Let us introduce the following notation:

$$\begin{aligned} i(x) &= \#\{v \in \widetilde{\Upsilon} : n_v = x\}; \\ i(x,k) &= \#\{v \in \widetilde{\Upsilon} : n_v \equiv x \bmod k\} = \sum_{\ell \in \mathbf{Z}} i(x + \ell k); \\ N &= \max\{n_v : v \in \widetilde{\Upsilon}\} : \\ g &= \sum_{v \in \widetilde{\Upsilon}} n_v = \sum_{x \geq 1} x i(x). \end{aligned}$$

Note for example that $i(x,N) = i(x)$ for all x such that $1 \leq x \leq N$.

The following is a consequence of Assumption 3.8.1:

LEMMA 3.8.2. *Suppose $r, s \in \mathbf{Z}$ are not divisible by k. If $\langle s \rangle \subseteq \langle r \rangle$ as subgroups of $\mathbf{Z}/k\mathbf{Z}$, $i(r,k) \leq i(s,k)$. Hence, if $\langle s \rangle = \langle r \rangle$, then $i(r,k) = i(s,k)$.*

LEMMA 3.8.3. *The x such that $i(x) \neq 0$, in other words the integers of the form n_v, are exactly the positive multiples of n_0 less than or equal to N.*

PROOF. It suffices to consider the case where $(\widetilde{\Upsilon}, \mathbf{n})$ is reduced and to show that the x such that $i(x) \neq 0$ are exactly the integers x such that $1 \leq x \leq N$. Let ℓ be the smallest positive integer such that $i(\ell) \neq 0$. Thus $i(t) = 0$ for $t < \ell$, and hence, by Lemma 3.8.2, $i(t) = 0$ for $N - \ell < t < N$. If $\ell \neq 1$, since $(\widetilde{\Upsilon}, \mathbf{n})$ is reduced, there exists an x with $i(x) \neq 0$ and with x not divisible by ℓ. Choose x to be the smallest such positive integer. In particular, $x > \ell$, and, by Lemma 3.8.2, $i(N - x) = i(x) \neq 0$. Since

$$-(x - \ell) = \ell - x \equiv N - x \bmod (N - \ell),$$

it follows that $i(x - \ell, N - \ell) \neq 0$. For $d \geq 1$,

$$d(N - \ell) + x - \ell \geq N - \ell + x - \ell > N - \ell,$$

and so if $t \equiv x - \ell \bmod (N - \ell)$ and $i(t) \neq 0$, then $t = x - \ell$. This says that $i(x - \ell) = i(x - \ell, N - \ell) \neq 0$. But $x - \ell < x$ and $\ell \nmid x - \ell$, contradicting the choice of x. Hence $\ell = 1$. Applying Lemma 3.8.2 with $k = N$ and $r = 1$, we see that $i(s) \geq i(1) \geq 1$ for all s with $1 \leq s \leq N$. □

LEMMA 3.8.4. *Suppose that $(\widetilde{\Upsilon}, \mathbf{n})$ is reduced. Then $\varphi(N) \leq 2$, where φ is the Euler φ-function, and hence $N \in \{1, 2, 3, 4, 6\}$.*

PROOF. Suppose that $1 \leq x \leq N - 1$ and that x is relatively prime to N. Then by Lemma 3.8.2 and Lemma 3.8.3, $i(x) = i(1) \geq 1$. Assume that $x \neq 1, N - 1$. Then

$$i(1) + 1 \leq i(1) + i(N) \leq i(1, N - 1) \leq i(x, N - 1) = i(x) = i(1),$$

a contradiction. Thus $x = 1$ or $x = N - 1$, and so $\varphi(N) \leq 2$. □

LEMMA 3.8.5. *Let $\ell > 1$ be a positive integer such that $i(\ell) \neq 0$. Then either $i(t\ell) = 0$ for $t > 1$ or $i(\ell) \geq 2$.*

PROOF. We may assume that $(\widetilde{\Upsilon}, \mathbf{n})$ is reduced and that $2\ell \leq N$. Since $\ell + 2 \leq 2\ell \leq N$, $i(1, \ell + 1) \geq 2$. First suppose that $\ell \geq 3$. By Lemma 3.8.2, $i(1, \ell + 1) = i(\ell, \ell + 1)$. Since $2\ell + 1 > 6 \geq N$, by Lemma 3.8.4, it follows that $i(\ell) = i(\ell, \ell + 1) = i(1, \ell + 1)$, which as we have just seen is at least 2. Now suppose that $\ell = 2$. Then

$N = 4$ or 6. If $N = 4$, then as before $2\ell + 1 > N$ and so $i(2) = i(2,3) \geq 2$. If $N = 6$, then $i(1,5) \geq 2$. But $i(1,5) = i(2,5) = i(2)$ and so again $i(2) \geq 2$. □

In case $(\widetilde{\Upsilon}, \mathbf{n})$ is reduced, it is easy to check that necessary and sufficient conditions on the integers $i(x)$ for the pair $(\widetilde{\Upsilon}, \mathbf{n})$ to satisfy Assumption 3.8.1 are as follows:

- For $N = 1$ resp. 2 there is no condition on the $i(x)$. In this case $g = i(1)$, resp. $g = i(1) + 2i(2)$.
- For $N = 3$, a necessary and sufficient condition is $i(1) = i(2)$. In this case $g = 3(i(1) + i(3))$.
- For $N = 4$, necessary and sufficient conditions are: $i(1) = i(3)$, $i(1) + i(4) = i(2)$. In this case $g = 6(i(1) + i(4))$.
- For $N = 6$, necessary and sufficient conditions are: $i(1) = i(5) = i(6)$, $i(2) = i(3) = i(4) = 2i(1)$. In this case $g = 30i(1)$.

Next we define

$$d_x = d_x(\widetilde{\Upsilon}, \mathbf{n}) = \#\{v \in \widetilde{\Upsilon} : x | n_v\} = \sum_{\ell \geq 1} i(\ell x).$$

Note that $d_x \neq 0$ if and only if $x | n_v$ for some v.

LEMMA 3.8.6.

$$\sum_{x \leq N} \varphi(x) d_x = g.$$

PROOF. Using the identity $\sum_{d|n} \varphi(d) = n$, we have

$$\sum_{x \leq N} \varphi(x) d_x = \sum_{v \in \widetilde{\Upsilon}} \sum_{d | n_v} \varphi(d) = \sum_{v \in \widetilde{\Upsilon}} n_v = g.$$

□

We come now to one form of the statement of clockwise symmetry:

THEOREM 3.8.7. *Suppose that the pair $(\widetilde{\Upsilon}, \mathbf{n})$ satisfies Assumption 3.8.1. For each $x \leq N$ such that $d_x \neq 0$, $x | 2g$. For each such $x \leq N$, and for each $r \leq x$ and relatively prime to x, consider the subset of the integers mod $2g$ given by*

$$J(x,r) = \left\{ \frac{2gr}{x} - d_x + 1, \frac{2gr}{x} - d_x + 3, \ldots, \frac{2gr}{x} + d_x - 3, \frac{2gr}{x} + d_x - 1 \right\}.$$

Thus $J(x,r)$ consists of d_x integers, centered at $2gr/x$ and with spacing 2. Then for distinct pairs $(x,r) \neq (y,s)$, the sets $J(x,r)$ and $J(y,s)$ are disjoint, and $\bigcup_{x,r} J(x,r) \subset \mathbf{Z}/2g\mathbf{Z}$ is either $\{0, 2, \ldots, 2g-2\}$ or $\{1, 3, \ldots, 2g-1\}$.

PROOF. Let us first show that it suffices to consider the the case where $(\widetilde{\Upsilon}, \mathbf{n})$ is reduced. For a general pair $(\widetilde{\Upsilon}, \mathbf{n})$, let $(\widetilde{\Upsilon}, \mathbf{n}')$ be the associated reduced pair. Thus $g = n_0 g'$. For each x, write $x = \ell m$, where $\ell = \gcd(x, n_0)$. Then $d_x = d'_m$, where $d'_m = d_m(\widetilde{\Upsilon}, n')$. Moreover, it is easy to see that $d_x = 0$ for all other x. An elementary argument shows that the set of rational numbers of the form $2g'r/m + a$, with $1 \leq r \leq m$ and r relatively prime to m, and $0 \leq a < n_0$, is exactly the set of rational numbers of the form $2n_0 g' s/x$, with $1 \leq s \leq x$ and s relatively prime to x and with $x = \ell m$, where $\ell = \gcd(x, n_0)$. Thus, $\bigcup_{x,r} J(x,r) \subset \mathbf{Z}/2g\mathbf{Z}$ for $(\widetilde{\Upsilon}, \mathbf{n})$ is invariant under translation by $2g'$ and the image of $\bigcup_{x,r} J(x,r)$ in $\mathbf{Z}/2g'\mathbf{Z}$ is the corresponding subset for the pair $(\widetilde{\Upsilon}, \mathbf{n}')$. Hence it suffices to consider the reduced case.

Let $\mathcal{F}_N = \{0/1, 1/N, 1/(N-1), \dots\}$ be the Farey sequence of rational numbers between 0 and 1 whose denominator is at most N, written in increasing order. We call integers x and y *adjacent* with respect to N if there exist r, s with $(r, x) = (s, y) = 1$ such that r/x and s/y are consecutive terms in $\mathcal{F}_N$. If r/x and s/y are consecutive terms in $\mathcal{F}_N$, then it is a standard fact that $sx - ry = 1$. The conclusions of Theorem 3.8.7 are easily seen to be equivalent to the following statement:

For all consecutive terms r/x and s/y in $\mathcal{F}_N$,

$$\frac{2gs}{y} = \frac{2gr}{x} + d_x + d_y.$$

Using the fact that $sx - ry = 1$, this condition is equivalent to:

For all integers x and y which are adjacent with respect to N,

$$g = \frac{xy}{2}(d_x + d_y).$$

Another way to write the conclusions of Theorem 3.8.7 is as follows: for all $x \leq N$ and $r \leq x$ with $(r, x) = 1$,

$$\frac{2gr}{x} = d_1 + \sum_{\substack{(y,t)=1 \\ 2gt/y < 2gr/x}} 2d_y + d_x.$$

By the symmetry $r \mapsto -r$ it is sufficient to check these conditions for $r/x \leq 1/2$. The case $r/x = 1/2$ follows from $\sum_{x \leq N} \varphi(x) d_x = g$. Thus, for $N \leq 4$, it is enough to check the first two conditions:

$$\begin{aligned} \frac{2g}{N} &= d_1 + d_N; \\ \frac{2g}{N-1} &= d_1 + 2d_N + d_{N-1}. \end{aligned}$$

Let us consider the first condition. By Assumption 3.8.1,

$$d_1 + d_N = \sum_{t=1}^{N-1} i(t) + 2i(N) = \sum_{\substack{t|N \\ t<N}} \varphi(N/t) i(t) + 2i(N).$$

Since $g = \sum_{t=1}^{N} ti(t) = \sum_{t=1}^{N-1} ti(t) + Ni(N)$, we see that it suffices to show that

$$\sum_{t=1}^{N-1} ti(t) = \sum_{\substack{t|N \\ t<N}} \frac{N}{2} \varphi(N/t) i(t).$$

On the other hand, by Assumption 3.8.1,

$$\sum_{t=1}^{N-1} ti(t) = \sum_{\substack{t|N \\ t<N}} \left(\sum_{\substack{s<N \\ \langle s \rangle = \langle t \rangle}} s \right) i(t).$$

The condition then follows from the elementary lemma:

LEMMA 3.8.8. *For every positive integer N, and every positive divisor t of N,*

$$\sum_{\substack{s<N \\ \langle s\rangle=\langle t\rangle}} s = \frac{N}{2}\varphi(N/t).$$

PROOF. Fix $t|N$. Then that

$$\sum_{\substack{s<N \\ \langle s\rangle=\langle t\rangle}} s = t \sum_{\substack{u<N/t \\ (u,N/t)=1}} u = t\frac{N}{2t}\varphi(N/t) = \frac{N}{2}\varphi(N/t),$$

where the second equality follows from

$$M\varphi(M) = \sum_{\substack{u<M \\ (u,M)=1}} M = \sum_{\substack{u<M \\ (u,M)=1}} (u + (M-u)) = 2 \sum_{\substack{u<M \\ (u,M)=1}} u.$$

□

This proves that the first condition holds under under Assumption 3.8.1. A very similar argument handles the second condition. The result follows for $N \leq 4$. The case $N = 6$ can be checked directly. □

REMARK 3.8.9. One can ask if, given a positive integer N there are collections of (not necessarily positive) integers $i(x)$, not satisfying Assumption 3.8.1, but such that the corresponding integers d_x satisfy the conclusions of Theorem 3.8.7. It is easy to see by the proof of Theorem 3.8.7 that, for $N \in \{1, 2, 3, 4, 6\}$, Theorem 3.8.7 is equivalent to Assumption 3.8.1. For $N = 5$, fixing a positive integer d and setting $i(1) = 2d, i(2) = 3d, i(3) = 3d, i(4) = 2d, i(5) = d$, the corresponding integers d_x satisfy Theorem 3.8.7 but of course the $i(x)$ cannot satisfy Assumption 3.8.1, by Lemma 3.8.4. These are in fact the only nonzero examples.

CHAPTER 4

c-pairs

Let C be an antisymmetric 2×2-matrix with entries in $\mathcal{C}G$. Then C is completely specified by $c = c_{12} = c_{21}^{-1}$. We consider ordered pairs of elements (x, y) in G satisfying $[x, y] = c$ and call such pairs *c-pairs*.

4.1. The rank zero case

Following our discussion of the structure of the moduli space of almost commuting N-tuples in G, our first task is to determine the set of rank zero c-pairs in G.

PROPOSITION 4.1.1. *Suppose that $c \in \mathcal{C}G$ is an element of order $k > 1$. Let (x, y) be a c-pair of rank zero in G. Then:*

1. *Both x and y are regular elements of G which are conjugate;*
2. *The group G is a product of r simple factors G_i, where each G_i is isomorphic to $SU(n_i)$ for some $i \geq 2$;*
3. *$c = c_1 \cdots c_r$ where c_i generates the center of G_i;*
4. *the subgroup of $G/\langle c \rangle$ generated by x, y is isomorphic to $(\mathbf{Z}/k\mathbf{Z})^2$ where k is the order of c;*
5. *All c-pairs in G are conjugate;*
6. *$Z(x, y) = \mathcal{C}G$.*

Conversely, if G is as in (2) and $c \in \mathcal{C}G$ is as in (3), then there is a rank zero c-pair in G.

PROOF. Let (x, y) be a c-pair in G. Conjugation by the element y normalizes the connected group $Z(x)$. Thus, by [**17**] II §2, since $Z(x, y)$ is finite, $Z(x)$ is a torus. Hence x is a regular element of G. By symmetry, y is also regular.

Recall that $A \subseteq i\mathfrak{t}$ is the alcove containing the origin associated to the set of simple roots Δ for Φ. By conjugation we can assume that $x \in T$, so that $T = Z(x)$. Conjugation by y normalizes $Z(x) = T$, and hence $y \in N_G(T)$. Finally, conjugation by an element of $N_G(T)$ makes $x = \exp(2\pi i\tilde{x})$ for a point $\tilde{x} \in A$. Since x is regular, $\tilde{x}$ is an interior point of A. Let $w \in W(T, G)$ be the Weyl element defined by conjugation by y. The relation $yxy^{-1} = xc^{-1}$ yields $w \cdot \tilde{x} - \xi = \tilde{x}$ for some $\xi \in i\mathfrak{t}$ such that $\exp(2\pi i\xi) = c^{-1}$. We denote by φ the affine linear map $v \mapsto w \cdot v - \xi$. The map φ normalizes the alcove structure for $W_{\text{aff}}(\Phi)$ and $\varphi(\tilde{x}) = \tilde{x}$, where $\tilde{x}$ is an interior point of A. Thus, $\varphi(A) = A$. It follows that $\xi \in A$ and, by Lemma 3.2.1, that φ is the action of c on A. This means that $w = w_c$. Since $T^{w_c} \subseteq Z(x, y)$, the fact that (x, y) is of rank zero implies that $\mathfrak{t}^{w_c}$ is a single point. Thus, Proposition 3.5.1 implies that G is a product of groups $\prod_{i=1}^{r} G_i$, where, for each i, the group G_i is isomorphic to $SU(n_i)$ for some $n_i > 1$, and c projects to a generator of $\mathcal{C}G_i$. The alcove A is a product of alcoves A_i for the simple factors G_i

of G. The unique fixed point of the c-action on A is the product of the barycenters of the A_i. Thus, $x = \exp(2\pi ib)$ where $b \in A$ is the product of the barycenters of the A_i.

Let $\langle x, y\rangle$ be the subgroup of G generated by x, y. In fact, it is a subgroup of $N_G(T)$ and $\langle x, y\rangle/(\langle x, y\rangle \cap T)$ is the cyclic group generated by $[y] = w_c \in W$. This element is of order k, the order of $c \in \mathcal{C}G$. Also, $x \in T$ is $\exp(2\pi ib)$, and so by inspection x has order k modulo $\langle c\rangle$. From this it is clear that $\langle x, y\rangle \subseteq G/\langle c\rangle$ is isomorphic to $(\mathbf{Z}/k\mathbf{Z})^2$.

Lastly, reversing the roles of x and y and replacing c by c^{-1} we see that y is also conjugate to the product of the barycenters of the A_i. Hence, x and y are conjugate in G.

Let (x', y') be any c-pair in G and let S be a maximal torus of $Z(x', y')$. By conjugation we can assume that $S = S_I$ for some subset $I \subseteq \Delta$. Then $c \in L_I$, and hence, by Lemma 3.4.1, $\Delta(c) \subseteq I$. Since G is a product of simple factors of type A and c projects to a generator of every factor, it follows from Proposition 3.5.1 that $\Delta(c) = \Delta$. Thus, $I = \Delta$ and consequently, S is trivial, which means that (x', y') is of rank zero. By what we proved above, we see that (x', y') is conjugate to a pair (x, y'') where $x = \exp(2\pi b)$ with b being the product of the barycenters of the A_i and $y'' \in w_cT$. But T operating by inner automorphism on w_cT is transitive since the component of the identity of T^{w_c} is trivial. [Proof: Consider the map μ defined by $t \mapsto tw_ct^{-1}$. The isotropy group of w_c is $\{t | w_c^{-1}tw_c = t\}$, which is finite.] Thus, all such pairs are conjugate by elements of T.

Suppose that $z \in Z(x, y)$. Since x is a regular element of T, $z \in T$ and $z = yzy^{-1} = w_cz$. Since G is a product of groups of type A and the image of c in each factor generates the center of that factor, it follows by inspection that $z \in \mathcal{C}G$.

Conversely, suppose that G is a product of simple groups isomorphic to $SU(n_i)$ and c projects to a generator of the center of each factor. Set x equal to the image under the exponential mapping of the product of the barycenters of the alcoves for the simple factors of G and take $y \in w_cT$. Then (x, y) is a c-pair of rank zero. □

COROLLARY 4.1.2. *If (x, y) is a rank zero c-pair in $SU(n)$, then (x, y) is conjugate to a pair of the form (x_0, y_0) where x_0 is the image under* $\exp(2\pi i\cdot)$ *of the barycenter of A and y_0 normalizes T and projects to w_c in $W(T, G)$. In particular, both x and y are regular elements and are conjugate in $SU(n)$. If n is odd, each of x and y has order n, which is the order of c. If n is even, then each of x and y has order $2n$, whereas the order of c is n. In fact, in this case $x^n = y^n = c^{n/2}$.*

PROOF. All of these statements were established in the course of the proof of the previous proposition, except the statements about the orders of x and y. They follow by inspection of the order of the image under $\exp(2\pi i\cdot)$ of the barycenter in a group of type A_{n-1}. □

REMARK 4.1.3. As explicit representatives for x and y, we can take:
(for n odd) $x = \text{diag}(\xi, \xi^2, \ldots, \xi^{n-1})$, $y = w_c\colon e_1 \mapsto e_2 \mapsto \cdots \mapsto e_n \mapsto e_1$, where ξ is a primitive n^{th} root of unity;
(for n even) $x = \text{diag}(\eta, \eta^3, \ldots, \eta^{2n-1})$, $y = w_c\colon e_1 \mapsto e_2 \mapsto \cdots \mapsto e_n \mapsto -e_1$, where η is a primitive $(2n)^{\text{th}}$ root of unity.

4.2. The general case

The next step is to determine the maximal torus for the centralizer of a c-pair (x, y).

PROPOSITION 4.2.1. *Let G be simple, let $c \in \mathcal{C}G$, and let $\mathbf{x} = (x, y)$ be a c-pair. We write*

$$c = \exp\left(2\pi i \sum_{a\in\Delta} \lambda_a a^\vee\right).$$

Then $I(\mathbf{x}) \subseteq \Delta$ is equal to

$$I_c = \{a \in \Delta | \lambda_a \notin \mathbf{Z}\}.$$

Thus any maximal torus of $Z(x, y)$ is conjugate in G to S_c. Finally, there is a unique rank zero c-pair up to conjugation in $L_c = DZ(S_c)$.

PROOF. Set $I = I(\mathbf{x})$. According to Corollary 2.3.3, $c \in L_I$ and there is a rank zero c-pair in L_I. Since L_I is simply connected, it follows from Proposition 4.1.1 that there are integers $n_1, \ldots, n_r$ such that L_I is isomorphic to $\prod_{i=1}^r SU(n_i)$ and under this isomorphism $c = \prod_{i=1}^r c_i$ where c_i generates the center of $SU(n_i)$. According to Proposition 3.5.1, no coefficient λ_a in

$$c = \exp(2\pi i \sum_{a\in I} \lambda_a a^\vee)$$

is an integer. It then follows from Lemma 3.4.1 that $I = I_c$, and hence that any maximal torus of $Z(x, y)$ is conjugate to S_c. It also follows from Proposition 4.1.1 that the c-pair in L_c is unique up to conjugation. □

COROLLARY 4.2.2. *Let $c \in \mathcal{C}G$ be given. Fix a c-pair (x_0, y_0) in L_c, and define a map $S_c \times S_c$ to the space of c-pairs in G by sending (s_1, s_2) to (s_1x_0, s_2y_0). This map factors to induce a homeomorphism from $(\overline{S}_c \times \overline{S}_c)/W(S_c, G)$ to the moduli space of conjugacy classes of c-pairs in G.*

PROOF. According to Proposition 4.1.1, the moduli space $\mathcal{M}^0_{L_I}(c)$ of c-pairs of rank zero in L_I is empty unless $I = I_c$ and the moduli space $\mathcal{M}^0_{L_{I_c}}(c)$ is a single point. Thus, the actions of $(F(S_c))^2$ and of $W(S_c, G)$ on $\mathcal{M}^0_{L_{I_c}}(c)$ are trivial. The result is now immediate from Corollary 2.3.3. □

PROOF OF THEOREM 1.3.1. Theorem 1.3.1 is an immediate consequence of Proposition 3.5.4 and Corollary 4.2.2.

Prior to Lemma 3.4.1, given c, we have defined a torus S_c and the group $DZ(S_c) = L_c$. There is also a description of the abstract group L_c in terms of the action of w_c:

PROPOSITION 4.2.3. *The group L_c is isomorphic to*

$$\prod_{\overline{a}\in\widetilde{\Delta}/\langle w_c\rangle} SU(n_{\overline{a}}),$$

where the $\overline{a}$ are the orbits of w_c acting on $\widetilde{\Delta}$ and $n_{\overline{a}}$ is the number of elements in $\overline{a}$.

PROOF. We know by Proposition 4.1.1 that L_c is isomorphic to a product of groups of the form $SU(n_i)$ and that, if c_i is the component of c in the ith factor, then c_i generates the center of $SU(n_i)$. The vector space $V = i\mathfrak{t}$ is a quotient of the vector space $\bigoplus_{a\in\tilde{\Delta}} \mathbf{R}\cdot a^\vee$ by the one-dimensional space spanned by $\sum_a g_a a^\vee$. The element w_c acts on $\bigoplus_{a\in\tilde{\Delta}} \mathbf{R}\cdot a^\vee$ by permuting the $a^\vee$ and is the identity on $\sum_a g_a a^\vee$. For each orbit $\overline{a}$, the subspace $\bigoplus_{b\in\overline{a}} \mathbf{R}\cdot a^\vee$ is w_c-invariant, and the eigenvalues of w_c on this subspace are $\zeta^i_{n_{\overline{a}}}$, $i = 0,\dots,n_{\overline{a}}-1$, where $\zeta_{n_{\overline{a}}}$ is a primitive $n_{\overline{a}}^{\text{th}}$ root of unity. On the other hand, w_c is conjugate to the product of the w_{c_j}, and the eigenvalues of w_{c_j} on the subspace of $\mathfrak{t}$ corresponding to the simple factor $SU(n_j)$ are $\zeta^i_{n_j}$, $i = 0,\dots,n_j-1$. The proposition follows by comparing the two forms for the set of eigenvalues. □

We conclude with a preliminary normal form for a c-pair; we shall give a more precise form in Chapter 6.

DEFINITION 4.2.4. A c-pair (x, y) is said to be *in weak normal form* (with respect to the maximal torus T and the alcove A) if $x \in T$, $x = \exp(2\pi i\tilde{x})$ for a point $\tilde{x} \in i\mathfrak{t}_c$ and $y \in N_G(T)$ projects to w_c in $W(T,G)$.

COROLLARY 4.2.5. *Every c-pair in G is conjugate to one in weak normal form.*

PROOF. By Corollary 4.2.2, after conjugation, we can assume that a maximal torus of $Z(x,y)$ is S_c. There is a rank zero c-pair (x_0, y_0) in $L_c = DZ(S_c)$ and $(s_1, s_2) \in S_c \times S_c$ such that $(x, y) = (s_1x_0, s_2y_0)$. The intersection $T \cap L_c$ is the maximal torus of L_c. According to Corollary 4.1.2 we can assume that $x_0 = \exp(2\pi i b)$ where b is the barycenter of an alcove A' of $i\mathfrak{t}_c$ and that the adjoint action of y_0 normalizes this subspace and has image in $W(L_c)$ equal to the Weyl part w of the action of c on the alcove A' for L_c. Thus, $x \in T$ and y normalizes T and projects to the image of $w \in W(L_c)$ in $W(G)$. Since this image is the Weyl part of the action of c on any alcove $A_1 \subset i\mathfrak{t}$ for G containing A', the result now follows by conjugating x and y by an element of $W(T,G)$ which sends A' into A. □

CHAPTER 5

Commuting triples

In this chapter G is simple and $c_{ij} = 1$ for $1 \le i < j \le 3$. In other words, we consider conjugacy classes of commuting triples (x, y, z) in G. We denote the moduli space of conjugacy classes of such triples by $\mathcal{T}_G$.

5.1. Commuting triples of rank zero

As usual, let A be the alcove T containing the origin corresponding to the set of simple roots Δ.

LEMMA 5.1.1. *Let (x, y, z) be a commuting triple in G of rank zero. Then x is conjugate in G to the image under* $\exp(2\pi i\cdot)$ *of a vertex v of the alcove A. Let $a \in \Delta$ be the root with the property that $\{a = 0\}$ defines the wall of A opposite v. Then $g_a = h_a$ and the order of x in G is g_a. For every $b \in \widetilde{\Delta} - \{a\}$ we have that $g_a \nmid g_b$. Each of y and z is conjugate in G to x and each has order equal to g_a. Conversely, let k be a positive integer such that $k | g_a$ for exactly one a. Then $k = g_a$ and there exists a commuting triple (x, y, z) in G of rank zero such that the order of x is k.*

PROOF. Let (x, y, z) have rank zero. By Lemma 2.2.3, $Z(x)$ is semi-simple and x is conjugate to the image under $\exp(2\pi i\cdot)$ of a vertex $v \in A$. Letting a be the simple root such that $\{a = 0\}$ defines the wall of A opposite v, we see that the order of x modulo $\mathcal{C}G$ is h_a. In particular, h_a divides the order of x.

The pair (y, z) is a commuting pair in $Z(x)$ of rank zero. Lift y, z to elements $\tilde{y}, \tilde{z}$ in the universal covering $\widetilde{Z}(x)$ of $Z(x)$. Then $(\tilde{y}, \tilde{z})$ is a rank zero c-pair for some $c \in \pi_1(Z(x)) \subseteq \mathcal{C}\widetilde{Z}(x)$. According to Lemma 3.1.2, the group $\pi_1(Z(x))$ is cyclic of order g_a and is generated by $c_0 = \exp(2\pi i\zeta)$, where

$$\zeta = -\frac{1}{g_a} \sum_{b \in \widetilde{\Delta} - \{a\}} g_b b^\vee.$$

By Proposition 4.1.1, the existence of a rank zero c-pair implies that c generates $\pi_1(Z(x))$ and that, if $\exp(2\pi i\zeta') = c$, then, in the expression of ζ' as a linear combination of the simple coroots, all coefficients are non-integral. Since c is a power of c_0, the same is true for ζ. This implies that, for each $b \in \widetilde{\Delta} - \{a\}$, the integer g_b is not divisible by g_a.

By Proposition 4.1.1, the elements $\tilde{y}$ and $\tilde{z}$ are conjugate in $\widetilde{Z}(x)$, and hence y and z are conjugate in $Z(x)$. It follows from the same proposition that the subgroup of $Z(x)$ generated by y, z is isomorphic to $(\mathbf{Z}/g_a\mathbf{Z})^2$. In particular, y and z have the same order g_a in G.

Interchanging the roles of x and y in this construction, we see that x and z are conjugate. Thus, x, y, z are all conjugate in G and hence all have the same order,

g_a. Since we have already shown that the order of x is divisible by h_a and since $g_a|h_a$, it follows that $g_a = h_a$.

To see the converse, suppose that $k|g_a$ for exactly one a. Then, by Lemma 3.8.5, $k = g_a$. Let x be the image under the exponential map of the vertex of the alcove opposite the face $\{a = 0\}$. By Theorem 3.6.1, the universal cover $\widetilde{Z}(x)$ is a product of groups of type A, and $Z(x) = \widetilde{Z}(x)/\langle\zeta\rangle$, where ζ has order k and projects to a generator of every factor. By Proposition 4.1.1, there is a rank zero ζ-pair $(\tilde{y}, \tilde{z})$ in $\widetilde{Z}(x)$. It suffices to take (y, z) to be the image in $Z(x)$ of $(\tilde{y}, \tilde{z})$. □

We define the *order of a commuting triple of rank zero* to be the common order of each of its elements.

COROLLARY 5.1.2. *Suppose that (x, y, z) is a rank zero commuting triple in G of order k. Then $\pi_1(Z(x))$ is a cyclic group of order k. If the image of $2\pi i\sum_{a\in\Delta}\lambda_a a^\vee$ under the exponential mapping lies in $\pi_1(Z(x)) \subseteq \mathcal{C}\widetilde{Z}(x)$ and generates this group, then $\lambda_a \notin \mathbf{Z}$ for every $a \in \Delta$.*

REMARK 5.1.3. If $G = SU(n+1)$, then there does not exist a rank zero commuting triple in G. This follows by Lemma 5.1.1, since all of the g_a are one in this case. Of course, it is elementary that every commuting N-tuple in $SU(n+1)$ is contained in a maximal torus, and hence can never be of rank zero.

REMARK 5.1.4. Suppose that (x, y, z) has rank zero and order $k > 1$. The lifts $\tilde{y}, \tilde{z}$ in the universal covering $\widetilde{Z}(x)$ form a c-pair for some $c \in \pi_1(Z(x)) \subseteq \mathcal{C}G$. It is easy to see directly that they generate a subgroup of $\widetilde{Z}(x)$ which modulo the center of $\widetilde{Z}(x)$ is isomorphic to $(\mathbf{Z}/k\mathbf{Z})^2$ and which meets the center in a subgroup of $\pi_1(Z(x))$. Thus, the elements y, z generate a subgroup of $Z(x)$ which is isomorphic to $(\mathbf{Z}/k\mathbf{Z})^2$ and has trivial intersection with the center of $Z(x)$. Since x lies in the center of $Z(x)$ and has order k, it follows that the group generated by x, y, z is isomorphic to $(\mathbf{Z}/k\mathbf{Z})^3$.

PROPOSITION 5.1.5. *There is a rank zero commuting triple of order k in G if and only if $k = g_a$ for exactly one a. In this case, there are exactly $\varphi(k)$ conjugacy classes of rank zero commuting triples of order k in G. If (x, y, z) has rank zero and order k, then the other conjugacy classes of such triples are represented by (x, y, z^ℓ) for $1 \le \ell < k$ and ℓ relatively prime to k.*

PROOF. The first statement follows from Lemma 5.1.1. Suppose that (x, y, z) is rank zero commuting triple of order k. According to Lemma 5.1.1, x is conjugate to the image under $\exp(2\pi i\cdot)$ of the vertex of the alcove opposite the face of A defined by $\{a = 0\}$ where a is the unique element of $\widetilde{\Delta}$ such that $k = g_a$. Then (y, z) is a rank zero commuting pair in $Z(x)$. Let $(\tilde{y}, \tilde{z})$ be a lift of (y, z) to the universal covering $\widetilde{Z}(x)$ of $Z(x)$. This is a rank zero c-pair for some c generating $\pi_1(Z(x)) \subseteq \mathcal{C}\widetilde{Z}(x)$. It follows that $\widetilde{Z}(x)$ is $\prod_{i=1}^r G_i$ with G_i isomorphic to $SU(n_i)$ for an integer $n_i|k$ and $c = c_1\cdots c_r$ where c_i generates the center of G_i. The element $c \in \pi_1(Z(x))$ depends only on the conjugacy class of (y, z) in $Z(x)$. By Proposition 4.1.1, c determines the conjugacy class of $(\tilde{y}, \tilde{z})$ in $\widetilde{Z}(x)$ and hence the conjugacy class of (y, z) in $Z(x)$. On the other hand, again by Proposition 4.1.1, for each $c' \in \pi_1(Z(x)) \subseteq \mathcal{C}\widetilde{Z}(x)$ there is a c'-pair $(\tilde{y}', \tilde{z}')$ in $\widetilde{Z}(x)$. Moreover, the c'-pair $(\tilde{y}', \tilde{z}')$ is of rank zero if and only if c' generates $\pi_1(Z(x))$, and in this case $(\tilde{y}', \tilde{z}')$ is unique up to conjugation in $\widetilde{Z}(x)$. The image (y', z') is a rank zero commuting pair in $Z(x)$. The group $\pi_1(Z(x))$ is cyclic of order k and hence has $\varphi(k)$ generators. This shows that there are exactly $\varphi(k)$ conjugacy classes of commuting pairs of rank zero in $Z(x)$, and hence $\varphi(k)$ conjugacy classes of rank zero commuting triples in G.

Clearly, if $[\tilde{y}, \tilde{z}] = c$ then $[\tilde{y}, \tilde{z}^\ell] = c^\ell$. This proves the last statement. □

5.2. A list of all simple groups with rank zero commuting triples

Suppose that G is simple and contains a rank zero commuting triple of order k. Then by Lemma 5.1.1, there is exactly one coroot integer g_a which is divisible by k, and in fact $g_a = k$. Conversely, if there is exactly one coroot integer g_a equal to k, then by Lemma 3.8.5, none of the other coroot integers is divisible by k and G contains a rank zero commuting triple of order k. Examining the coroot integers on the Dynkin diagrams of the simple groups, one sees that the following are the only possibilities:

(1) $k = 1$: G is the trivial group.
(2) $k = 2$: G is of type D_4, of type B_3, or of type G_2.
(3) $k = 3$: G is of type E_6 or F_4.
(4) $k = 4$: G is of type E_7.
(5) $k = 5$ or $k = 6$: G is of type E_8.

5.3. Action of the outer automorphism group of G

PROPOSITION 5.3.1. *Let $\mathcal{T}_G^0$ be the space of conjugacy classes of commuting triples of rank zero in G. Then the action of every automorphism σ of G on $\mathcal{T}_G^0$ is trivial.*

PROOF. Let (x, y, z) be a rank zero commuting triple in G. Let k be the order of (x, y, z). There is a unique $a \in \widetilde{\Delta}$ such that $k | g_a$. After conjugation we can assume that x is the exponential of the vertex $\tilde{x}$ of the alcove A opposite the face $\{a = 0\}$. After composing σ with a suitable inner automorphism, we can assume that σ normalizes Δ. The action of σ on the set of simple roots preserves the integers $\{g_b\}_{b \in \widetilde{\Delta}}$. Hence $\sigma(a) = a$ and therefore $\sigma(\tilde{x}) = \tilde{x}$ and $\sigma(x) = x$. Thus σ acts on $\pi_1(Z(x))$, and by Corollary 3.1.3, since σ preserves the coroot integers g_b, the action is trivial. Hence σ acts trivially on the conjugacy class of (y, z) in $Z(x)$, by Proposition 4.1.1, and therefore on the conjugacy class of (x, y, z) in G. □

5.4. Action of the center of G

There is an action of $(\mathcal{C}G)^3$ on the space of conjugacy classes of commuting triples defined by $(\gamma_1, \gamma_2, \gamma_3) \cdot (x, y, z) = (\gamma_1 x, \gamma_2 y, \gamma_3 z)$. Clearly, $Z(x, y, z) = Z(\gamma_1 x, \gamma_2 y, \gamma_3 z)$ so that this action preserves the subspace of conjugacy classes of commuting triples of rank zero.

PROPOSITION 5.4.1. *The induced action of $(\mathcal{C}G)^3$ on $\mathcal{T}_G^0$ is trivial.*

PROOF. Let (x, y, z) be a rank zero commuting triple of order $k > 1$. First consider the action of $\mathcal{C}G$ on x. We can assume that x is the image under $\exp(2\pi i \cdot)$ of the vertex of the alcove A opposite the face $\{a = 0\}$ where $a \in \widetilde{\Delta}$ is the unique element with $k = g_a$. For any $\gamma \in \mathcal{C}G$ let w_γ be the Weyl element which is the linear part of the action of c on A. The Weyl element w_γ normalizes $\widetilde{\Delta}$ and, according to Section 3.2, preserves the g_b in the sense that $g_{w_\gamma \cdot b} = g_b$. This implies that $w_\gamma \cdot a = a$. Thus, the affine automorphism φ_γ of A fixes x. By Equation 3.1, this means that if $h \in N_G(T)$ projects to $w_\gamma \in W$, then $hxh^{-1} = x\gamma^{-1}$. Thus the triples $(x\gamma, y, z)$ and $(x, {}^h y, {}^h z)$ are conjugate.

CLAIM 5.4.2. *Conjugation by h normalizes $Z(x)$ and induces the identity automorphism of its fundamental group.*

PROOF. Since $Z(x) = Z(x\gamma)$ the first statement is clear. By Corollary 3.1.3, the map $\zeta \mapsto \exp(2\pi i \zeta)$ identifies the subgroup of $(Q^\vee \otimes \mathbf{Q})/Q^\vee$ generated by $\zeta = (-1/g_a)\sum_{b \in \widetilde{\Delta} - \{a\}} g_b b^\vee$ with the fundamental group of $Z(x)$. The element w_γ normalizes the set $\widetilde{\Delta} - \{a\} \subset i\mathfrak{t}^*$ and the subset $\widetilde{\Delta}^\vee - \{a^\vee\} \subset i\mathfrak{t}$. Since $g_{w_\gamma b} = g_b$, for $b \in \widetilde{\Delta}$, it is clear that w_γ fixes $\zeta \in i\mathfrak{t}$, and hence acts trivially on $\pi_1(Z(x))$. □

It follows immediately from the claim that $({}^h y, {}^h z)$ lifts to a c-pair in $\widetilde{Z}(x)$, where $c = [\tilde{y}, \tilde{z}]$ for any two lifts of y, z to $\widetilde{Z}(x)$. By Proposition 4.1.1, $({}^h y, {}^h z)$ is conjugate in $Z(x)$ to (y, z). Hence $(x\gamma, y, z)$ is conjugate to (x, y, z), and so the action of $\mathcal{C}G$ on the first factor of commuting triples induces the trivial action on the space of conjugacy classes of commuting triples of rank zero. The situation is completely symmetric in x, y, z and it then follows that the action of $\mathcal{C}G$ on the space of conjugacy classes of commuting triples of rank zero is trivial. □

5.5. The general case

Let Φ be a root system on a vector space V. Let the coroot integers be $\{g_a\}$. If $k \geq 1$ divides at least one of the g_a, we set $\widetilde{I}(k) \subseteq \widetilde{\Delta} = \{a \in \widetilde{\Delta} : k \nmid g_a\}$. Let $f(k)$ be the face of A which is the intersection of all the walls of A defined by the $a \in \widetilde{I}(k)$ and let

$$V(k) = \bigcap_{a \in \widetilde{I}(k)} \operatorname{Ker} a.$$

Of course $V(k)$ is the linear space parallel to $f(k)$. By convention, if $k = 1$ then we let $V(k) = V$. In all other cases $V(k)$ is a proper subspace of V. In the case that $\Phi = \Phi(T, G)$ so that $V = i\mathfrak{t}$ we define $\mathfrak{t}(k)$ so that $i(\mathfrak{t}(k)) = V(k)$. In this case we let $S(k)$ be the subtorus of T whose Lie algebra is $\mathfrak{t}(k)$ and let $L(k)$ be the derived group of $Z(S(k))$.

PROPOSITION 5.5.1. *Let (x, y, z) be a commuting triple. Then there is a unique positive integer $k \geq 1$ dividing at least one of the g_a with the following properties:*

(1) *$S(k)$ is conjugate to a maximal torus of $Z(x, y, z)$.*
(2) *After conjugation, we can find a decomposition $(x, y, z) = (s_1 x_0, s_2 y_0, s_3 z_0)$ where $s_i \in S(k)$ and (x_0, y_0, z_0) is a commuting triple of rank zero and of order k in $L(k)$.*
(3) *The element x is conjugate to an element of the form* $\exp(2\pi i \tilde{x})$ *for some element $\tilde{x} \in f(k)$.*

We call k the order *of the commuting triple (x, y, z).*

PROOF. Let (x, y, z) be a commuting triple, let S be a maximal torus of $Z(x, y, z)$ and let $\mathfrak{s} = \operatorname{Lie}(S)$. By Theorem 2.3.1, there is a commuting triple (x_0, y_0, z_0) of rank zero in $L = DZ(S)$ such that $x \in S \cdot x_0$. Let k be the order of (x_0, y_0, z_0). If $k = 1$, then L is trivial. In this case, $x, y, z \in S$ and hence they are all contained in a maximal torus. Thus S itself is a maximal torus for G. Conversely, if x, y, z are contained in a maximal torus T, then T is a maximal torus for $Z(x, y, z)$ and hence L is trivial. For the rest of the proof, we assume that L is not trivial, or equivalently that S is not a maximal torus.

CLAIM 5.5.2. *L is a simply connected, simple group, not of type A. If G is not simply laced, then L is not simply laced.*

PROOF. Let $L = \prod_i L_i$ be the decomposition of L into its simple factors. Since $L = DZ(S)$, L is simply connected and the Dynkin diagram of L is identified with a subdiagram of the Dynkin diagram of G. Thus, at most one of the components L_i is not of type A. Since L has a commuting triple of rank zero, it follows that each of the L_i has such a triple. By Remark 5.1.3 this implies that no L_i is of type A. Thus, L is simple. The last statement is clear since, if G is not simply laced, then $D(G)$ is a chain, and hence every simply laced subdiagram is of type A. □

CLAIM 5.5.3. *For a generic $x' \in S \cdot x_0$, $DZ(x') = Z_L(x_0)$. In particular, $Z_L(x_0)$ is semi-simple.*

PROOF. Suppose that x' is generic in $S \cdot x_0$. The roots in $DZ(x')$ are the roots that are integral on x'. Since x' is generic, a root is integral on x' if and only if it is integral on S and on x_0. This shows that the roots of G integral on x' are exactly the roots of L integral on x_0 and hence that $DZ(x') = DZ_L(x_0)$. Since (x_0, y_0, z_0) is a rank zero commuting triple in L, the center of $Z_L(x_0)$ is finite. Since $Z_L(x_0)$ is connected, it is semi-simple and equal to its own derived group. Thus $DZ(x') = DZ_L(x_0) = Z_L(x_0)$. □

Let $x' \in S \cdot x_0$ be generic. After conjugation, we can assume that $x' = \exp(2\pi i\tilde{x})$ for a point in $\tilde{x} \in A$. We let $\widetilde{I}(x') \subseteq \widetilde{\Delta}$ be the subset consisting of all the roots which are one on x'. The subset $\widetilde{I}(x')$ forms a set of simple roots for the root system of $DZ(x')$ with respect to the maximal torus $T \cap DZ(x')$.

It now follows from Corollary 5.1.2 that $\pi_1(Z_L(x_0))$ is a cyclic group of order k, and hence $\pi_1(DZ(x'))$ is cyclic of order k. By Corollary 3.1.3, this means that k divides g_a for every $a \in \widetilde{\Delta} - \widetilde{I}(x')$. That is to say $\widetilde{I}(k) \subseteq \widetilde{I}(x')$. Hence $x' = \exp(2\pi i\tilde{x}')$ for some $\tilde{x}' \in f(k)$. Since this is true for a generic $x' \in S \cdot x_0$, the same conclusion holds for every point $x_1 \in S \cdot x_0$, and in particular for x_0.

Since (x_0, y_0, z_0) is a rank zero triple in L, it follows from Corollary 5.1.2 that if I is any set of simple roots for the root system $Z_L(x_0)$ and if $\zeta = \sum_{a \in I} \lambda_a a^\vee$ has the property that $\exp(2\pi i\zeta)$ is a central element in $\widetilde{Z}_L(x_0)$ generating $\pi_1(Z_L(x_0))$, then no λ_a is an integer. But according to Corollary 3.1.3, $\widetilde{I}(x')$ is a set of simple roots for the root system $DZ(x') = Z_L(x_0)$ and, if

$$\chi = -\frac{1}{k} \sum_{a \in \widetilde{I}(x')} g_a a^\vee,$$

then $\exp(2\pi i\chi)$ generates the fundamental group of $DZ(x') = Z_L(x_0)$. It follows that $k \nmid g_a$ for every $a \in \widetilde{I}(x')$. This implies that $\widetilde{I}(x') \subseteq \widetilde{I}(k)$, and hence that $\widetilde{I}(x') = \widetilde{I}(k)$.

Hence $x' = \exp(2\pi i\tilde{x}')$ for a point $\tilde{x}'$ contained in the interior of the face $f(k)$ of A and $DZ(x')$ has $\widetilde{I}(k)$ as a set of simple roots. Consequently, $Z_L(x_0)$ has $\widetilde{I}(k)$ as a set of simple roots. Since $Z_L(x_0)$ is semi-simple, $\mathrm{Lie}(Z_L(x_0)) \cap \mathfrak{t} = \mathrm{Lie}(L) \cap \mathfrak{t}$.

Thus, $\mathfrak{s} = \text{Lie}(S)$ is the intersection of the kernels of the $a \in \widetilde{I}(k)$. It follows that $\mathfrak{s} = \mathfrak{t}(k)$ and $L = L(k)$. □

We now establish a converse to Proposition 5.5.1:

PROPOSITION 5.5.4. *Suppose that $k \geq 1$ is a positive integer dividing at least one of the g_a. Then there exists a commuting triple of order k in G.*

PROOF. We begin with the following result about root systems:

PROPOSITION 5.5.5. *Let Φ be a reduced and irreducible root system on a vector space V, and suppose that $k > 1$ is an integer such that $k|g_a$ for some a. Define $\widetilde{I}(k)$ and $V(k)$ as before, and let $\Phi(k)$ be the set of all roots which annihilate $V(k)$. Then $\Phi(k)$ is an irreducible root system. Moreover, if the coroot integers for $\Phi(k)$ are of the form $m_b, b \in \widetilde{\Delta}(\Phi(k))$, then $k|m_b$ for exactly one b, and in this case $k = m_b$.*

PROOF. We may assume that Φ is the root system of the group G. Thus there is a torus $S(k)$ corresponding to $V(k) = i\mathfrak{t}(k)$. Let $L(k) = Z(S(k))$. Then $\Phi(k)$ is the set of roots for $L(k)$, and in particular it is a root system. Let $Q^\vee_{L(k)}$ be the sublattice of $Q^\vee$ generated by the coroots of $L(k)$. Then $Q^\vee_{L(k)}$ is a primitive sublattice of $Q^\vee$, by Lemma 2.1.1. Let $Q^\vee_{\widetilde{I}(k)}$ be the lattice spanned by $\widetilde{I}(k)$. Then $Q^\vee_{\widetilde{I}(k)}$ is a sublattice of $Q^\vee_{L(k)}$. By Lemma 3.1.2, $\text{Tor}(Q^\vee/Q^\vee_{\widetilde{I}(k)})$ is a cyclic group of order equal to the gcd of the g_a such that $k|g_a$, and by Corollary 3.6.3, this gcd is k. Since $Q^\vee_{L(k)}$ is a primitive sublattice of $Q^\vee$, it follows that $Q^\vee_{L(k)}/Q^\vee_{\widetilde{I}(k)}$ is also cyclic of order k.

Next, we have the following description of the root system $\Phi(k)$:

CLAIM 5.5.6. *Let Φ^+ be the set of positive roots for Φ corresponding to Δ and let $I(k) = \widetilde{I}(k) \cap \Delta$. Then $\Phi^+(k) = \Phi^+ \cap \Phi(k)$ is a set of positive roots for $\Phi(k)$. Let $\Delta(k)$ be the corresponding set of simple roots. Then $\Delta(k) = I(k) \cup \{b\}$ for some root $b \in \Phi(k)$. The root system $\Phi(k)$ is irreducible and the highest root d for Φ is also a highest root for $\Phi(k)$.*

PROOF. Choose any $\tilde{p}$ contained in the interior of A. Then the roots of $\Phi(k)$ which are positive on $\tilde{p}$ are exactly those in $\Phi^+(k)$. Thus, $\Phi^+(k)$ is a set of positive roots with respect to some set of simple roots of $\Phi(k)$. Clearly, the elements of $I(k) \subseteq \Phi(k)$ are positive roots. Since none of these can be written as a non-trivial linear combination of positive roots of Φ, *a fortiori* none of these can be written as a non-trivial linear combination of positive roots of $\Phi(k)$. Thus $I(k)$ is a subset of the set of simple roots $\Delta(k)$ determined by $\Phi^+(k)$. Since $g_{\tilde{a}} = 1$, we have $\tilde{a} \in \widetilde{I}(k)$, and hence the cardinality of $I(k)$ is one less than the dimension of the span of $\Phi(k)$. Thus, there is a root $b \in \Phi^+(k)$ with the property that $\Delta(k) = I(k) \cup \{b\}$.

Let d be the highest root of Φ with respect to the positive roots Φ^+. Since $\tilde{a} \in \widetilde{I}(k)$, $d = -\tilde{a} \in \Phi^+(k)$. Write $d^\vee = \sum_{a \in I(k)} m_a a^\vee + m_b b^\vee$. Since $\{a^\vee, a \in I(k)\} \cup \{b^\vee\}$ is a basis for $Q^\vee_{L(k)}$ and $\{a^\vee, a \in I(k)\} \cup \{d^\vee\}$ is a basis for $Q^\vee_{\widetilde{I}(k)}$, it follows that $Q^\vee_{L(k)}/Q^\vee_{\widetilde{I}(k)}$ is cyclic of order m_b. Thus $m_b = k$. Since $k \nmid g_a$ for all $a \in I(k)$, it follows that $m_a \neq 0$ for all $a \in I(k)$. This proves that all the coefficients is this expression are non-trivial, and hence $\Phi(k)$ is irreducible. Since the sum of d and any positive root in Φ is not a root of Φ, it follows that d is the highest root of $\Phi(k)$ with respect to the set of simple roots $\Delta(k)$. □

Returning to the proof of Proposition 5.5.5, we see that we have proved that $\Phi(k)$ is irreducible and that $k = m_b$ and $k \nmid m_a$ for $a \neq b$. This completes the proof of Proposition 5.5.5. □

Finally, let us finish the proof of Proposition 5.5.4. Since $L(k)$ is a simple group and k is equal to exactly one of the coroot integers of $L(k)$, it follows by Proposition 5.1.5 that $L(k)$ contains a commuting rank zero triple of order k. Of course, such a triple will also be a commuting triple of order k in G. □

THEOREM 5.5.7. *Let G be simple. Let $k \geq 1$ be an integer.*

(1) *If (x, y, z) is a commuting triple of order k in G, then k divides at least one of the coroot integers g_a and $S(k)$ is conjugate to a maximal torus for $Z(x, y, z)$.*
(2) *The order is a conjugacy class invariant and defines a locally constant function on $\mathcal{T}_G$.*
(3) *If k divides at least one of the g_a, there are exactly $\varphi(k)$ components of $\mathcal{T}_G$ consisting of conjugacy classes of commuting triples of order k, where φ is the Euler φ-function. Given a component X of $\mathcal{T}_G$, let $d_X = \frac{1}{3}\dim X + 1$. Then*

$$\sum_X d_X = g.$$

(4) *Each component consisting of commuting triples of order k in G is homeomorphic to*

$$\left(\overline{S}(k) \times \overline{S}(k) \times \overline{S}(k)\right)/W(S(k), G).$$

PROOF. The first statement follows from Proposition 5.5.1. Clearly, the order is a conjugacy class invariant, and it is locally constant on $\mathcal{T}_G$ by Corollary 2.3.2. Now suppose that k divides at least one of the g_a. By Proposition 5.5.4, there is a commuting triple $\mathbf{x}$ of order k in G. By Part (1), we can assume after conjugation that $S(k)$ is a maximal torus of $Z(\mathbf{x})$. By Part (2) of Proposition 5.5.1, there is a rank zero commuting triple of order k in $L(k)$. It then follows from Lemma 5.1.5 there are exactly $\varphi(k)$ conjugacy classes of commuting triples of rank zero in $L(k)$. By Proposition 5.4.1, the center of $L(k)$ acts trivially on the space of conjugacy classes of commuting triples in $L(k)$ or, in the notation of Corollary 2.3.3, the group F^3 acts trivially on the space of conjugacy classes of commuting triples in $L(k)$. By Proposition 5.3.1, the Weyl group of $W(S(k), G)$ acts trivially on the set of conjugacy classes of rank zero commuting triples in $L(k)$. Corollary 2.3.3 now implies that there are exactly $\varphi(k)$ components of $\mathcal{T}_G$ of triples of order k, and each of these components is homeomorphic to

$$(\overline{S}(k) \times \overline{S}(k) \times \overline{S}(k))/W(S(k), G),$$

proving the first sentence in Part (3) and Part (4) of the theorem. Let X be a component of $\mathcal{T}_G$ of order k. By Part (4),

$$\frac{1}{3}\dim X + 1 = \dim S(k) + 1.$$

It follows directly from the definition of $S(k)$ that $\dim S(k) + 1$ is equal to the number of a such that $k|g_a$. The second statement of Part (3) then follows from the first statement of Part (3) and Lemma 3.8.6. □

The first four parts of Theorem 1.4.1 are contained in the statement of Theorem 5.5.7. We shall prove the last item of Theorem 1.4.1 in Chapter 9.

REMARK 5.5.8. Assume that $G \neq L(k)$, in other words that the c-triple (x, y, z) has positive rank. We have defined $\overline{S}(k)$ to be $S(k)/(S(k) \cap L(k))$. Here $S(k) \cap L(k) \subseteq \mathcal{C}L(k)$, and is easily checked to be $\pi_k(2\pi i Q^\vee)/2\pi i Q^\vee_{L(k)}$, where $Q^\vee_{L(k)}$ is the coroot lattice of $L(k)$ and π_k is orthogonal projection onto the real vector space spanned by $2\pi i Q^\vee_{L(k)}$. Using this remark, it is not difficult to check that $S(k) \cap L(k) = \mathcal{C}L(k)$ except for the case where G is of type D_n for $n > 4$ and $k = 2$, so that $L(2)$ is of type D_4. In this case, $S(2) \cap L(2)$ has order 2.

CHAPTER 6

Some results on diagram automorphisms and associated root systems

Throughout this chapter, Φ denotes a reduced but not necessarily irreducible root system on the vector space V with a basis Δ a set of simple roots. Let $\mathcal{A}$ be the decomposition of V into alcoves determined by the set of affine walls $\mathcal{W}$ associated to Φ. Suppose that τ is a group of affine isometries of V normalizing the alcove decomposition $\mathcal{A}$. Suppose that A_0 is an an alcove such that $\tau(A_0) = A_0$. After conjugating by an element of the affine Weyl group group, we can assume that $A_0 = A$ is the alcove associated to Δ. If Φ is irreducible, then A is a simplex and every group τ of affine isometries of A fixes the barycenter of A, which is an interior point of A. In general, the group τ fixes the product of the barycenters of the factors of A. Let ℓ be the associated group of linear isomorphisms of V and let V^ℓ be the fixed subspace of ℓ. Clearly, the group ℓ normalizes $\widetilde{\Delta}$ and defines a group of diagram automorphisms of $\widetilde{D}(\Phi)$. Conversely, every such group of diagram automorphisms leads to a group of affine automorphisms τ as above. We denote by $\widetilde{\Delta}/\ell$ the quotient set. Note that ℓ acts on the set Φ of all roots as well.

The purpose of this chapter is to study the set of nonzero restrictions of elements of Φ to V^ℓ. We show that this set forms a root system $\Phi^{\mathrm{res}}(\ell)$ whose Weyl group is the Weyl group of V^ℓ in W, and explicitly identify the inverse coroots. There are two other closely related root systems with the same Weyl group which we also study. Related results, in a more general context, have been given in [**7**] and [**6**].

6.1. A chamber structure and a Coxeter group on the fixed subspace

LEMMA 6.1.1. (1) *No wall W of $\mathcal{W}$ contains the fixed subspace V^τ of τ.*

(2) *If W is a wall of $\mathcal{W}$ meeting V^τ, then the intersection $W^\tau = W \cap V^\tau$ is a codimension-one affine subspace of V^τ.*

(3) *The walls W^τ divide V^τ into compact convex subsets with nonempty interior. We denote this collection of subsets by $\mathcal{A}^\tau$.*

(4) *The elements of $\mathcal{A}^\tau$ are exactly the subsets $V^\tau \cap A'$, where A' is an alcove of $\mathcal{A}$ such that $\tau(A') = A'$.*

PROOF. Since V^τ contains an interior point of an alcove, no wall of $\mathcal{W}$ can contain V^τ. The second and third statements are now clear. As to the last, if A' is an alcove of $\mathcal{A}$ normalized by τ, then $A' \cap V^\tau$ contains an interior point of A', and hence $A' \cap V^\tau$ is the closure of its interior in V^τ. Clearly, $A' \cap V^\tau \in \mathcal{A}^\tau$. Conversely, let $B \in \mathcal{A}^\tau$. Since B contains a non-empty open subset of V^τ, it contains a element of $V - \bigcup_{W \in \mathcal{W}} W$. This shows that B is contained in a unique alcove A' of $\mathcal{A}$. Clearly, $B = V^\tau \cap A'$ and $\tau(A') = A'$. $\square$

LEMMA 6.1.2. *Two elements of $\widetilde{\Delta}$ have the same restriction to V^ℓ if and only if they lie in the same ℓ-orbit. If $a \in \widetilde{\Delta}$, then $a|V^\ell \neq 0$.*

PROOF. Clearly, it suffices to establish this result in the case that Φ is irreducible, so that A is a simplex. Suppose that $a, a' \in \widetilde{\Delta}$ are in the same ℓ-orbit. Then their restrictions to V^ℓ are equal. Furthermore, the walls W_a and $W_{a'}$ of A determined by a and a' are in the same τ-orbit. This means that $W_a \cap V^\tau = W_{a'} \cap V^\tau$.

Since the restrictions of the walls of A to V^τ cut out a compact convex body, there must be at least $\dim(V^\tau) + 1$ distinct and non-parallel walls. But this is exactly the cardinality of $\widetilde{\Delta}/\ell$. Hence, it follows that distinct ℓ-orbits in $\widetilde{\Delta}$ cut out distinct and non-parallel walls in V^τ, and hence have distinct, nonempty restrictions to V^ℓ. □

Note: it is not in general true that a general $a \in \Phi$ has nonzero restriction to V^ℓ, or that, if two elements of Φ have the same (nonzero) restriction to V^ℓ, then they lie in the same ℓ-orbit. See for example Lemma 6.6.1 below.

LEMMA 6.1.3. (1) *The group τ normalizes the affine Weyl group $W_{\mathrm{aff}}(\Phi)$.*
(2) *The centralizer $Z_{W_{\mathrm{aff}}(\Phi)}(\tau)$ is equal to the normalizer $N_{W_{\mathrm{aff}}(\Phi)}(V^\tau)$ of V^τ in $W_{\mathrm{aff}}(\Phi)$ and acts simply transitively on $\mathcal{A}^\tau$.*

PROOF. Since τ normalizes $\mathcal{A}$, it normalizes the group generated by reflections in the walls of $\mathcal{W}$, i.e., the affine Weyl group $W_{\mathrm{aff}}(\Phi)$. To establish (2), clearly $Z_{W_{\mathrm{aff}}(\Phi)}(\tau) \subseteq N_{W_{\mathrm{aff}}(\Phi)}(V^\tau)$. Conversely, suppose that $g \in N_{W_{\mathrm{aff}}(\Phi)}(V^\tau)$. Then $B = gA$ is an element of $\mathcal{A}$ meeting V^τ in an interior point. Thus for all $f \in \tau$, $f^{-1}gfA = B = gA$. By Part 1, $f^{-1}gf \in W_{\mathrm{aff}}(\Phi)$ and hence $f^{-1}gf = g$. Since this is true for all $f \in \tau$, $g \in Z_{W_{\mathrm{aff}}(\Phi)}(\tau)$. The final statement is now clear since $W_{\mathrm{aff}}(\Phi)$ acts simply transitively on the set of all alcoves. □

LEMMA 6.1.4. *For each wall $W \cap V^\tau$ there is an element of $Z_{W_{\mathrm{aff}}(\Phi)}(\tau)$ which is a geometric reflection in this wall. The group $Z_{W_{\mathrm{aff}}(\Phi)}(\tau)$ is generated by the reflections in the walls of any given element of $\mathcal{A}^\tau$. Thus, $Z_{W_{\mathrm{aff}}(\Phi)}(\tau)$ is an affine Coxeter group with fundamental domain $B = A \cap V^\tau$.*

PROOF. The wall $W \cap V^\tau$ is a common wall between two alcoves B_1, B_2 of $\mathcal{A}^\tau$. Let A_1, A_2 be the τ-invariant alcoves of $\mathcal{A}$ containing B_1, B_2. Let $g \in Z_{W_{\mathrm{aff}}(\Phi)}(\tau)$ be the unique element carrying A_1 to A_2. Then g is a product of reflections about walls separating A_1 and A_2, and hence it is the identity on $A_1 \cap A_2$ and *a fortiori* on $B_1 \cap B_2$. Since $B_1 \cap B_2$ contains a nonempty open subset of $W \cap V^\tau$, $g|V^\tau$ is an isometry fixing $W \cap V^\tau$ and sending B_1 to B_2. It is then the reflection in $W \cap V^\tau$. □

PROPOSITION 6.1.5. *Let τ be a group of affine isometries of V normalizing the alcove decomposition associated to Φ. Suppose that τ normalizes the alcove A. Then there is a point $v \in B = A \cap V^\tau$ which is a vertex of the alcove decomposition $\mathcal{A}^\tau$ so that, using v to identify V^τ with V^ℓ, there is a uniquely determined reduced root system Φ^τ on V^ℓ whose alcove structure is $\mathcal{A}^\tau$. The affine Weyl group of Φ^τ is $Z_{W_{\mathrm{aff}}(\Phi)}(\tau)$. The root system Φ^τ is irreducible if Φ is irreducible. There is a set of extended simple roots $\widetilde{\Delta}^\tau$ for Φ^τ and a bijection $\iota\colon \widetilde{\Delta}/\ell \to \widetilde{\Delta}^\tau$ such that the restriction mapping $\widetilde{\Delta}/\ell \to \left(V^\ell\right)^*$ sends $\overline{a} \in \widetilde{\Delta}/\ell$ to a positive multiple of $\iota(\overline{a}) \in \widetilde{\Delta}^\tau$.*

PROOF. We may assume that Φ is irreducible. By Lemma 6.1.4, there is a Coxeter group with $A \cap V^\tau$ as fundamental domain. By the general classification result for such Coxeter groups [**4**], it follows that this Coxeter group is isomorphic to the affine Weyl group of a (reduced) root system Φ^τ. That is to say there is a linear structure on V^τ compatible with its given affine structure, such that under this identification the Coxeter group becomes the affine Weyl group of a root system. Of course this linear structure is determined by choosing a point $v \in V^\tau$ to identify V^τ with V^ℓ. The point v must be a vertex of an alcove. In fact, we can choose it to be a vertex of the alcove $A \cap V^\tau$. When we do this, the restricted roots defining the walls of $A \cap V^\tau$ become the set of extended simple roots. By Lemma 6.1.2 these roots are exactly the walls of the restrictions of the orbits $\widetilde{\Delta}/\ell$. The proof of Lemma 6.1.2 shows that $A \cap V^\tau$ has $\dim V^\tau + 1$ walls. Hence $A \cap V^\tau$ is a simplex, so that Φ^τ is irreducible. The last statement follows since a root is determined up to a multiple by the wall it defines, and it is easy to see in this case that the multiple must be positive. □

COROLLARY 6.1.6. *The Weyl group $W(\Phi^\tau)$ is the group of isometries of V^ℓ generated by the reflections in $\{\overline{a}\}_{\overline{a}\in\widetilde{\Delta}/\ell}$.*

As a first application to the study of c-pairs we have the following:

LEMMA 6.1.7. *Let (x,y) be a c-pair in G. Then up to conjugation we can assume that $x = \exp(2\pi i\tilde{x})$, where $\tilde{x}$ lies in the fixed set A^c of the alcove A, and $y \in N_G(T)$ is an element projecting to $w_c \in W(T,G) = W$.*

PROOF. By Corollary 4.2.5, we may assume that x is the image under $\exp(2\pi i\cdot)$ of a point $\tilde{x} \in it^c$ and that $y \in N_G(T)$ projects to w_c in W. By Proposition 6.1.5 applied to the affine automorphism φ_c of Section 3.2, there is a $\gamma \in W_{\mathrm{aff}}(G)$, commuting with φ_c, such that $\gamma\cdot\tilde{x} \in A^c$. Let $h \in N_G(T)$ project to the element of W which is the linear part of γ. Then hxh^{-1}, hyh^{-1} satisfy the conclusions of the lemma. □

DEFINITION 6.1.8. A c-pair (x,y) is said to be in *normal form* (with respect to the maximal torus T and the alcove A) if $x \in T$, x is the image under $\exp(2\pi i\cdot)$ of a point $\tilde{x} \in A^c$ and $y \in N_G(T)$ projects to w_c in $W(T,G)$. By the above lemma, every c-pair is conjugate to one in normal form.

6.2. The restricted root system and the projection root system

In this section we keep the notation of Section 6.1. Let $\widetilde{D}(\Phi)$ be the extended diagram of Φ. The action of τ permutes the connected components of $\widetilde{D}(\Phi)$. If D is a connected component of $\widetilde{D}(\Phi)$, then the stabilizer τ_D acts as a group of diagram automorphisms of D. Moreover, given $a \in D$, the τ-orbit of a is a disjoint union of mutually perpendicular copies of the τ_D-orbit of a. Likewise, V is an orthogonal direct sum of subspaces V_D indexed by the components of $\widetilde{D}(\Phi)$, and V^ℓ is an orthogonal direct sum over the τ-orbits of the set of components of $\widetilde{D}(\Phi)$, of the fixed spaces $V_D^{\ell_D}$, where ℓ_D is the linear group corresponding to τ_D. Thus, in the proofs that follow, we will be able to reduce to the case where Φ is irreducible.

LEMMA 6.2.1. *Let $\mathcal{O}$ be an orbit of ℓ acting on $\widetilde{\Delta}$. Then exactly one of the following holds:*

(1) *$\mathcal{O}$ is a union of components D of $\widetilde{D}(\Phi)$ of type A and the stabilizer τ_D of a component D of $\mathcal{O}$ acts transitively on D. In this case, if V_D is the subspace of V corresponding to a component D of $\mathcal{O}$ and ℓ_D is the linear group corresponding to the stabilizer τ_D of D in τ, then ${V_D}^{\ell_D} = \{0\}$.*
(2) *For all $a, b \in \mathcal{O}$ with $a \neq b$, a and b are orthogonal. In this case we say that $\mathcal{O}$ is a* ordinary *orbit.*
(3) *$\mathcal{O} = \coprod_i \mathcal{O}_i$, where each $\mathcal{O}_i$ is of cardinality 2, say $\mathcal{O}_i = \{a_{i,1}, a_{i,2}\}$. Furthermore, $a_{i,1}$ and $a_{i,2}$ have the same lengths and $n(a_{i,1}, a_{i,2}) = -1$. Lastly, elements from distinct $\mathcal{O}_i$ are orthogonal. We say that $\mathcal{O}$ is an* exceptional *orbit and that each of the $\mathcal{O}_i$ is an* exceptional pair. *If moreover Φ is irreducible, not of type A and τ is cyclic, then there is at most one exceptional orbit and, if it exists, it has exactly two elements.*

PROOF. By the remarks before the statement of the lemma, it suffices to consider the case where Φ is irreducible. If Φ is of type A, then the corresponding group of diagram automorphisms is a cyclic group or a dihedral group and the lemma follows by inspection. Otherwise, $\widetilde{D}(\Phi)$ is a connected, contractible diagram which is not a single orbit. Let D be the proper subdiagram of $\widetilde{D}(\Phi)$ defined by $\mathcal{O}$ and suppose that D is a union of k connected subdiagrams D_i. Then the D_i are all isomorphic and the stabilizer of D_i in ℓ acts transitively on the nodes of D_i. It follows that each D_i is of type A_1 or A_2. The first case corresponds to an ordinary orbit, the second to an exceptional orbit. The final statement of Case 3 follows easily from the Lefschetz fixed point formula. □

We remark that Case 1 above occurs if and only if τ_D contains a rotation of order $n+1$, or $n+1 = 2k$ is even, τ_D contains a rotation of order k, and an involution with no fixed points.

LEMMA 6.2.2. *Let $B = A \cap V^\tau$. Let W be the wall of B corresponding to the orbit $\mathcal{O} \subseteq \widetilde{\Delta}$. Suppose that $a \in \Phi$ and that the wall $W_a \cap V^\tau$ is equal to W. Then either $\pm a \in \mathcal{O}$ or $\mathcal{O}$ is exceptional and $\pm a = a_1 + a_2$ for an exceptional pair $\{a_1, a_2\}$ in $\mathcal{O}$.*

PROOF. Choose a point $\tilde{x} \in B \cap W$ which is contained in no other wall. By Lemma 6.1.2, the set of roots in $\widetilde{\Delta}$ which take integral values on $\tilde{x}$ is exactly $\mathcal{O}$. By Lemma 3.1.1, since $\tilde{x} \in A$, the set of roots a in Φ such that a is integral on $\tilde{x}$ is a root system with simple roots equal to $\mathcal{O}$. If $\mathcal{O}$ is ordinary, then this root system exactly is a product of root systems of type A_1 and one of $a, -a$ is contained in $\mathcal{O}$. Otherwise $\mathcal{O}$ is exceptional and this root system is a product of root systems of type A_2. Every root of this system, up to sign, is either in $\mathcal{O}$ or is the sum of an exceptional pair in $\mathcal{O}$. □

Using the previous lemma and Weyl invariance, we can extend the description of the possible orbit types from the set of orbits of extended roots to all orbits.

COROLLARY 6.2.3. *Let $a \in \Phi$ and suppose that $a|V^\ell \neq 0$. Let $\mathcal{O}$ be the orbit of a. Then either the elements of $\mathcal{O}$ are mutually orthogonal or $\mathcal{O}$ is a disjoint union of mutually orthogonal subsets $\mathcal{O}_i$ where each $\mathcal{O}_i = \{a_{i,1}, a_{i,2}\}$ and where $a_{i,1}$ and $a_{i,2}$ have the same length and $n(a_1, a_2) = -1$.*

PROOF. Since $a|V^\ell \neq 0$, the wall W_a corresponding to a meets V^τ in a hyperplane. Thus there is $g \in Z_{W_{\rm aff}(\Phi)}(\tau)$ such that $g \cdot W_a$ defines a wall of B. Suppose that $g \cdot W_a$ corresponds to the orbit $\mathcal{O}'$ of simple roots. If w is the linear part of g, then w commutes with ℓ and hence sends ℓ-orbits to ℓ-orbits. The result now follows from Lemma 6.2.2. □

In the first case of the corollary, we call $\mathcal{O}$ *ordinary* and in the second case we call $\mathcal{O}$ *exceptional* and the subsets $\mathcal{O}_i$ *exceptional pairs.*

For $a \in \Phi$, let $\overline{a}$ be the ℓ-orbit of a. For $a \in \Phi$, $a|V^\ell$ is a linear form, depending only on the orbit $\overline{a}$. Given an orbit $\overline{a}$, let $n_{\overline{a}}$ be the number of elements of $\overline{a}$. Define the *restricted roots* $\Phi^{\mathrm{res}}(\ell) \subseteq (V^\ell)^*$ to be the set of nonzero linear maps of the form $a|V^\ell$ for $a \in \Phi$. Note that distinct orbits may define the same restricted root, although this does not happen for orbits contained in $\widetilde{\Delta}$.

For an orbit $\overline{a}$ in Φ/ℓ we define $\epsilon(\overline{a}) = 1$ if $\overline{a}$ is ordinary and $\epsilon(\overline{a}) = 2$ if $\overline{a}$ is exceptional. Now we define the coroots inverse to the elements of $\Phi^{\mathrm{res}}(\ell)$ as follows: For each $u \in \Phi^{\mathrm{res}}(\ell)$ we choose $a \in \Phi$ such that $a|V^\ell = u$ and define the inverse coroot

$$u^\vee = \epsilon(\overline{a}) \sum_{a' \in \overline{a}} (a')^\vee. \tag{6.1}$$

CLAIM 6.2.4. *$u^\vee$ is independent of the choice of $a \in \Phi$ restricting to give u and $u^\vee = \epsilon(\overline{a}) n_{\overline{a}} \pi(a^\vee)$, where π is orthogonal projection $V \to V^\ell$.*

PROOF. Elements $a, b \in \Phi$ restrict to give the same root in $\Phi^{\mathrm{res}}(\ell)$ if and only if $\pi(a^\vee) = \pi(b^\vee)$. Clearly, $\epsilon(\overline{a}) \sum_{a' \in \overline{a}} (a')^\vee$ is a positive real multiple of $\pi(a^\vee)$ and by Corollary 6.2.3 $\langle \pi(a^\vee), \epsilon(\overline{a}) \sum_{a' \in \overline{a}} (a')^\vee \rangle = 2$. The proves the first statement. The second follows from the fact that $\pi(a^\vee) = (1/n_{\overline{a}}) \sum_{a' \in \overline{a}} (a')^\vee$. □

PROPOSITION 6.2.5. *The set $\Phi^{\mathrm{res}}(\ell)$ is a possibly nonreduced root system in V^ℓ. It is irreducible if Φ is irreducible. For $u \in \Phi^{\mathrm{res}}(\ell)$, the coroot inverse to u is $u^\vee$ as given in Equation 6.1. The Weyl group of $\Phi^{\mathrm{res}}(\ell)$ is equal to $W(\Phi^\tau)$.*

We call $\Phi^{\mathrm{res}}(\ell)$ the *restricted root system.*

PROOF. Clearly the $\{\overline{a} : a \in \Phi\}$ span the dual space to V^ℓ and hence the roots of $\Phi^{\mathrm{res}}(\ell)$ span this space. As we saw in the above claim, for any $u \in \Phi^{\mathrm{res}}(\ell)$ the inner product $\langle u, u^\vee \rangle$ is 2. It is clear from the definitions that for $u, v \in \Phi^{\mathrm{res}}(\ell)$ we have $\langle u, v^\vee \rangle \in \mathbf{Z}$. Thus it suffices to show that, for all $u, v \in \Phi^{\mathrm{res}}(\ell)$, we have

$$r_u(v) = v - \langle v, u^\vee \rangle u \in \Phi^{\mathrm{res}}(\ell).$$

We fix $a \in \Phi$, resp. $b \in \Phi$, such that restriction of a, resp. b to V^ℓ is u, resp. v. Then

$$\begin{aligned} r_u(v) &= v - \langle v, u^\vee \rangle u = v - \langle v, \epsilon(\overline{a}) \sum_{a' \in \overline{a}} (a')^\vee \rangle u \\ &= v - \langle b, \epsilon(a) \sum_{a' \in \overline{a}} (a')^\vee \rangle u \\ &= \left(b - \langle b, \epsilon(a) \sum_{a' \in \overline{a}} (a')^\vee \rangle a \right) |V^\ell \end{aligned}$$

First assume that $\overline{a}$ is an ordinary orbit. Since $a'|V^\ell = a|V^\ell$ for all $a' \in \overline{a}$ and since $\epsilon(\overline{a}) = 1$, we have

$$r_u(v) = \left(b - \sum_{a' \in \overline{a}} \langle b, (a')^\vee \rangle a'\right) |V^\ell.$$

Suppose that $\overline{a} = \{a_1, \dots, a_n\}$ where the a_i are pairwise distinct. Since the $a_i \in \overline{a}$ are mutually orthogonal, this last equation can be rewritten as

$$r_u(v) = r_{a_1} \circ r_{a_2} \circ \cdots \circ r_{a_n}(b)|V^\ell.$$

(Notice that the r_{a_i} commute, so that the composition is independent of the ordering.) Clearly, then, $r_u(v) \in \Phi^{\mathrm{res}}(\ell)$.

In case $\overline{a}$ is an exceptional orbit $\mathcal{O} = \coprod_{i=1}^t \mathcal{O}_i$ with the $\mathcal{O}_i = \{a_{i,1}, a_{i,2}\}$ being exceptional pairs we have

$$r_u(v) = (b - 2\langle b, \sum_i a_{i,1}^\vee + a_{i,2}^\vee \rangle a_{i,1})|V^\ell = (b - \langle b, \sum_i a_{i,1}^\vee + a_{i,2}^\vee \rangle (a_{i,1} + a_{i,2}))|V^\ell,$$

since $a_{i,1}$ and $a_{i,2}$ have the same restriction to ℓ. In this case, $c_i = a_{i,1} + a_{i,2}$ is a root and $c_i^\vee = a_{i,1}^\vee + a_{i,2}^\vee$ since $a_{i,1}$, $a_{i,2}$, and c_i have the same length. Thus

$$r_u(v) = r_{c_1} \circ \cdots \circ r_{c_t}(b)|V^\ell.$$

(Here, the c_i are orthogonal, so that the order of the reflections is again irrevelant.) This proves that $r_u(v) \in \Phi^{\mathrm{res}}(\ell)$ in this case also.

The walls in V^ℓ defined by the elements of $\Phi^{\mathrm{res}}(\ell)$ are the same as the walls of Φ^τ, viewed as linear hyperplanes. Thus, the Weyl groups are the same. Finally, if Φ^τ is irreducible, we cannot divide the set of walls for Φ^τ into two nonempty, mutually orthogonal subsets. Thus the same is true for $\Phi^{\mathrm{res}}(\ell)$, and hence $\Phi^{\mathrm{res}}(\ell)$ is irreducible. □

For $a \in \Phi$ such that $a|V^\ell \neq 0$, the element $\pi(a^\vee)$ is given by the formula

$$\pi(a^\vee) = \frac{1}{n_{\overline{a}}} \sum_{b \in \overline{a}} b^\vee.$$

Define $\Phi^{\mathrm{proj}}(\ell)^\vee \subseteq V^\ell$ by

$$\Phi^{\mathrm{proj}}(\ell)^\vee = \{\, \pi(a^\vee) : a \in \Phi \,\} - \{0\}.$$

Clearly, $\pi(a^\vee) = 0$ if and only if $\overline{a} = 0$ as a linear form.

We now define a second root system in V^ℓ as follows. Its coroots will be the set $\Phi^{\mathrm{proj}}(\ell)^\vee$. Given $a \in \Phi$ such that $\pi(a^\vee) \neq 0$, we define the root inverse to $\pi(a^\vee) \in \Phi^{\mathrm{proj}}(\ell)^\vee$ to be $\epsilon(\overline{a}) n_{\overline{a}} a|V^\ell$. As before, this is independent of the choice of a lift of $\pi(a^\vee)$ to $a \in \Phi^\vee$. Let $\Phi^{\mathrm{proj}}(\ell) \subseteq V^\ell$ be the set of all such elements.

Dually to the above results, we have:

PROPOSITION 6.2.6. *$\Phi^{\mathrm{proj}}(\ell)$ is a possibly nonreduced root system in V^ℓ. It is irreducible if Φ is irreducible. The coroot inverse to $\epsilon(\overline{a}) n_{\overline{a}} a|V^\ell$ is $\pi(a^\vee)$. The coroot lattice of $\Phi^{\mathrm{proj}}(\ell)$ is $\pi(Q^\vee)$. The Weyl group of $\Phi^{\mathrm{proj}}(\ell)$ is the same as that of $\Phi^{\mathrm{res}}(\ell)$ and hence as that of Φ^τ.*

We call $\Phi^{\mathrm{proj}}(\ell)$ the *projection root system.*

The set of walls defined by $\Phi^{\mathrm{proj}}(\ell)$ is equal to the set of walls defined by $\Phi^{\mathrm{res}}(\ell)$, and thus the two systems have the same Weyl groups. In general, however, there is no one-to-one correspondence between $\Phi^{\mathrm{proj}}(\ell)$ and $\Phi^{\mathrm{res}}(\ell)$. Of course, if $\tau = \mathrm{Id}$, then $\Phi^{\mathrm{proj}}(\ell) = \Phi^{\mathrm{res}}(\ell) = \Phi$. On the other hand, there are examples where one of the systems is reduced and the other is non-reduced. Note however that the set $\widetilde{\Delta}/\ell$ injects into both $\Phi^{\mathrm{res}}(\ell)$ and $\Phi^{\mathrm{proj}}(\ell)$. Moreover, if Φ is simply laced, we can say the following:

LEMMA 6.2.7. *Suppose that Φ is simply laced. Then, using the Weyl invariant inner product to identify V and V^*, $\Phi^{\mathrm{proj}}(\ell)$ is the inverse system to $\Phi^{\mathrm{res}}(\ell)$.*

PROOF. In case Φ is simply laced, the inner product identifies a with $a^\vee$ and $a|V^\ell$ with $\pi(a^\vee) \in V^\ell$. Thus the roots of $\Phi^{\mathrm{res}}(\ell)$ are identified with the coroots in $\Phi^{\mathrm{proj}}(\ell)$. □

6.3. Generalized Cartan matrices for $\Phi^{\mathrm{res}}(\ell)$ and $\Phi^{\mathrm{proj}}(\ell)^\vee$

In this chapter we suppose that Φ is irreducible. We identify $\widetilde{\Delta}/\ell$ with its image in V^ℓ and hence for $a \in \widetilde{\Delta}$ we write $\overline{a}$ both for an orbit in $\widetilde{\Delta}$ and for an element of $\Phi^{\mathrm{res}}(\ell)$. Let $\pi(a^\vee)$ be the element of $\Phi^{\mathrm{proj}}(\ell)^\vee$ corresponding to $a^\vee$, but continue to denote the corresponding orbit of $\widetilde{\Delta}$ by $\overline{a}$.

Since $\widetilde{\Delta}/\ell \subseteq V^\ell$ are elements of the root system $\Phi^{\mathrm{res}}(\ell)$, and since from Equation 6.1, $n(\overline{a},\overline{b}) \leq 0$ for $\overline{a} \neq \overline{b}$, $\overline{a},\overline{b} \in \widetilde{\Delta}/\ell$, the numbers $n(\overline{a},\overline{b})$ form a generalized Cartan matrix. Two elements $\overline{a}$ and $\overline{b}$ are orthogonal if and only if their orbits span orthogonal subspaces of V. Since Φ is irreducible, it follows that this generalized Cartan matrix is indecomposable.

Let $d = \dim V^\ell$. Then the cardinalities of $\widetilde{\Delta}/\ell$ and of $\widetilde{\Delta}^\vee/\ell$ are both $d+1$. Since $\widetilde{\Delta}/\ell$, resp. $\widetilde{\Delta}^\vee/\ell$, spans $(V^\ell)^*$, resp. V^ℓ, there is a single linear relation among its elements. We claim that the one relation has positive integral coefficients. The root integers h_a only depend on the orbit $\overline{a}$. Define $h_{\overline{a}} = n_{\overline{a}} h_a$ for any choice of $a \in \overline{a}$. The relation $\sum_{a\in\widetilde{\Delta}} h_a a = 0$ leads to a relation

$$\sum_{\overline{a}\in\widetilde{\Delta}/\ell} h_{\overline{a}}\overline{a} = 0,$$

Similarly, the relation for the projection coroots is:

$$\sum_{\overline{a}} g_{\overline{a}}\pi(a^\vee) = 0, \tag{6.2}$$

where $g_{\overline{a}} = n_{\overline{a}} g_a$. Thus, the generalized Cartan matrices determined by $n(\overline{a},\overline{b})$ and by $n(\pi(a^\vee),\pi(b^\vee))$ are of affine type. It is not in general true that $\widetilde{\Delta}/\ell$, resp. $\widetilde{\Delta}^\vee/\ell$ is an extended set of simple roots resp. coroots for $\Phi^{\mathrm{res}}(\ell)$ resp. $\Phi^{\mathrm{proj}}(\ell)$, cf. Proposition 6.6.4.

There are then a corresponding affine diagrams, which we denote by $\widetilde{D}(\widetilde{\Delta}/\ell)$ and $\widetilde{D}(\Delta^\vee/\ell)$. While these affine Dynkin diagrams will be different in general, the associated Coxeter graphs will be the same. (Here the Coxeter graph of a generalized Dynkin diagram is obtained by keeping bonds and their multiplicities but forgetting the arrows.)

Our goal now will be to work out explicitly the Cartan integers for the coroots $\Phi^{\mathrm{proj}}(\ell)^\vee$ inverse to the projection root system. We begin with a graph-theoretic lemma:

LEMMA 6.3.1. *Let D be a finite tree, and let ℓ be a group of automorphisms of D. If v_1 and v_2 are two vertices of D which are connected by an edge, then either* $\mathrm{Stab}(v_1) \subseteq \mathrm{Stab}(v_2)$ *or* $\mathrm{Stab}(v_2) \subseteq \mathrm{Stab}(v_1)$.

PROOF. First we claim that there is fixed point for the action of ℓ on the topological space $|D|$ associated to D. The proof is by induction on the number of vertices. Clearly ℓ has a fixed point if there are 1 or 2 vertices. Otherwise, let $D' \subset D$ be the subgraph obtained by deleting the leaves. Then D' is a nonempty contractible proper subgraph on which ℓ acts, so by induction there is a fixed point of the action of ℓ on D' and hence on D.

Choose a point p fixed by ℓ. If p is an interior point of an edge e whose boundary is $\{v_1, v_2\}$, then it is easy to see that $\mathrm{Stab}(v_1) = \mathrm{Stab}(v_2)$ is the set of $g \in \ell$ such that $g|e = \mathrm{Id}$. Assume that we are not in this case. There is a unique path Γ in D joining p to v_1. Possibly after switching v_1 and v_2, we can assume that v_2 does not lie on this path. Hence the unique path Γ' from p to v_2 is the union of Γ with the edge connecting v_1 and v_2. If $g \in \ell$ fixes v_2, then $g(\Gamma') = \Gamma'$. Since $g(p) = p$, $g|\Gamma' = \mathrm{Id}$. Thus $g(v_1) = v_1$. It follows that $\mathrm{Stab}(v_2) \subseteq \mathrm{Stab}(v_1)$. ☐

PROPOSITION 6.3.2. *Suppose that Φ is not of type A. Let $\pi(a^\vee), \pi(b^\vee) \in \Phi^{\mathrm{proj}}(\ell)^\vee$, and let $n(\pi(a^\vee), \pi(b^\vee))$ be the corresponding Cartan integer. Then:*

(1) *If every element of $\overline{a}$ is orthogonal to every element of $\overline{b}$, then*
$$n(\pi(a^\vee), \pi(b^\vee)) = 0.$$

(2) *If there exist $a \in \overline{a}$ and $b \in \overline{b}$ such that $a^\vee$ and $b^\vee$ are not orthogonal, then either* $\mathrm{Stab}(a^\vee) \subseteq \mathrm{Stab}(b^\vee)$ *or* $\mathrm{Stab}(b^\vee) \subseteq \mathrm{Stab}(a^\vee)$. *If* $\mathrm{Stab}(a^\vee) \subseteq \mathrm{Stab}(b^\vee)$, *then*
$$n(\pi(a^\vee), \pi(b^\vee)) = \epsilon(\overline{b}) n(a^\vee, b^\vee).$$

(3) *If there exist $a \in \overline{a}$ and $b \in \overline{b}$ such that $a^\vee$ and $b^\vee$ are not orthogonal and* $\mathrm{Stab}(b^\vee) \subseteq \mathrm{Stab}(a^\vee)$, *then*
$$n(\pi(a^\vee), \pi(b^\vee)) = \epsilon(\overline{b}) \frac{n_{\overline{b}}}{n_{\overline{a}}} n(a^\vee, b^\vee).$$

PROOF. Let $a^\vee, b^\vee \in \widetilde{\Delta}^\vee$ have the property that $n(a_i^\vee, b_j^\vee) = 0$ for all $a_i \in \overline{a}$ and all $b_j \in \overline{b}$. The two subspaces of V spanned by the $a_i^\vee$ such that $a_i \in \overline{a}$, resp. the $b_j^\vee$ such that $b_j \in \overline{b}$, are orthogonal, and hence $\pi(a^\vee)$ and $\pi(b^\vee)$ in V^ℓ are also orthogonal. Thus $n(\pi(a^\vee), \pi(b^\vee)) = 0$ and there is no bond between $\pi(a^\vee)$ and $\pi(b^\vee)$ in the affine diagram associated with the generalized Cartan matrix of these elements.

Suppose that there exist $a \in \overline{a}$ and $b \in \overline{b}$ such that $n(a^\vee, b^\vee) \neq 0$. By the previous lemma, either $\mathrm{Stab}(a^\vee) \subseteq \mathrm{Stab}(b^\vee)$ or $\mathrm{Stab}(b^\vee) \subseteq \mathrm{Stab}(a^\vee)$. Since the root inverse to $\pi(b^\vee)$ is $\epsilon(\overline{b}) n_{\overline{b}} b | V^\ell$ we have

$$\begin{aligned} n(\pi(a^\vee), \pi(b^\vee)) &= \langle \pi(a^\vee), \epsilon(\overline{b}) n_{\overline{b}} \overline{b} \rangle \\ &= \epsilon(\overline{b}) \langle \pi(a^\vee), \sum_{b' \in \overline{b}} b' \rangle = \epsilon(\overline{b}) \langle a^\vee, \sum_{b' \in \overline{b}} b' \rangle \\ &= \epsilon(\overline{b}) \sum_{b' \in \overline{b}} \langle a^\vee, b' \rangle = \epsilon(\overline{b}) \sum_{b' \in \overline{b}} n(a^\vee, (b')^\vee). \end{aligned}$$

If $\mathrm{Stab}(a^\vee) \subseteq \mathrm{Stab}(b^\vee)$, we see that

$$\sum_{b' \in \bar{b}} n(a^\vee, (b')^\vee) = n(a^\vee, b^\vee),$$

and $n(\pi(a^\vee), \pi(b^\vee)) = \epsilon(\bar{b}) n(a^\vee, b^\vee)$ in this case. On the other hand if $\mathrm{Stab}(b^\vee) \subseteq \mathrm{Stab}(a^\vee)$, then

$$\sum_{b' \in \bar{b}} n(a^\vee, (b')^\vee) = \frac{n_{\bar{b}}}{n_{\bar{a}}} n(a^\vee, b^\vee),$$

and thus

$$n(\pi(a^\vee), \pi(b^\vee)) = \epsilon(\bar{b}) \frac{n_{\bar{b}}}{n_{\bar{a}}} n(a^\vee, b^\vee).$$

□

Similar results handle the case where Φ is of type A:

PROPOSITION 6.3.3. *Suppose that Φ is of type A, and that $V^\ell \neq 0$. Then the Cartan integers are given by the same formulas as Proposition 6.3.2 except in the case where Φ is of type A_{2k-1}, ℓ contains a rotation of order k and an involution fixing two vertices. In this case, the quotient coroot diagram is of type $\widetilde{A}_1$.*

PROOF. Since ℓ is dihedral, the stabilizer of an element has either one or two elements. The only case not covered by Proposition 6.3.2 is the case where there exist two non-orthogonal coroots $a^\vee$ and $b^\vee$ such that $\mathrm{Stab}(a^\vee)$ and $\mathrm{Stab}(b^\vee)$ are both nontrivial. In this case, the product of the two nontrivial elements is a rotation which either has order $n+1$, if n is even, or $k = (n+1)/2$, if n is odd. In the first case, $V^\ell = \{0\}$, and in the second case either $V^\ell = \{0\}$ or Φ is of type A_{2k-1}, the rotation subgroup ℓ' of ℓ has order exactly k, and there is an involution in ℓ fixing two vertices. In this case, $V^\ell = V^{\ell'}$, and the quotient coroot diagram is of type $\widetilde{A}_1$. □

Note that the affine diagrams associated to the Cartan integers we have calculated here agree with the diagrams given in Definition 1.6.1 in the introduction.

Let V' be a subspace of V, and let $W(V', \Phi) = N_{W(\Phi)}(V')/Z_{W(\Phi)}(V')$, a finite group. In case $V = i\mathfrak{t}$ and S is a subtorus of T with $\mathrm{Lie}(S) = \mathfrak{s}$, then $W(i\mathfrak{s}, \Phi(T, G)) = W(\mathfrak{s}, G)$ as defined in the introduction.

PROPOSITION 6.3.4. *Let Φ be a root system with τ and ℓ as above. Suppose that, for every component D of $\widetilde{D}(\Phi)$ which is of type A, the stabilizer ℓ_D in ℓ of D is either trivial or is not a cyclic group of rotations of D. Then the Weyl group $W(V^\ell, \Phi)$ is identified with the Weyl group of $\Phi^{\mathrm{res}}(\ell)$ or equivalently with the Weyl group of $\Phi^{\mathrm{proj}}(\ell)$.*

PROOF. It suffices to consider the case when Φ is irreducible. Clearly, the result holds if ℓ is the trivial group. Thus, we assume that $\ell \neq \mathrm{Id}$. We have seen that we can realize the elements of the Weyl group of $\Phi^{\mathrm{res}}(\ell)$ or of $\Phi^{\mathrm{proj}}(\ell)$ as elements of the Weyl group $W(\Phi)$ normalizing V^ℓ. Thus there is a homomorphism from the Weyl group of $\Phi^{\mathrm{res}}(\ell)$ to $W(V^\ell, \Phi)$, and since the Weyl group of $\Phi^{\mathrm{res}}(\ell)$ acts faithfully on V^ℓ, this homomorphism is injective. We must show that its image is all of $W(V^\ell, \Phi)$.

Given $w \in W(V^\ell, \Phi)$, represent w by an element of $W(\Phi)$ which normalizes V^ℓ. Since w permutes the set $\{\bar{a}\}$ of restricted roots and preserves the inner product on

V^ℓ, it defines an automorphism of the root system $\Phi^{\mathrm{proj}}(\ell)$. We claim that $\Phi^{\mathrm{proj}}(\ell)$ is either non-simply laced or A_1. Assuming this, since every automorphism of a root system which is either non-simply laced or A_1 is given by a Weyl element, it follows that w is given by an element of the Weyl group of $\Phi^{\mathrm{proj}}(\ell)$, or equivalently of $\Phi^{\mathrm{res}}(\ell)$.

Note that that $\Phi^{\mathrm{proj}}(\ell)$ is non-simply laced if there is an exceptional orbit. Thus we may assume that there are no exceptional orbits. First assume that Φ is not of type A. Since the diagram $\widetilde{D}(\Phi)$ is contractible and there are no exceptional orbits, there is a vertex a fixed by ℓ, and hence $n_{\overline{a}} = 1$. Since ℓ is nontrivial, not all of the $n_{\overline{b}}$ are 1. In particular there must exist two coroots $a^\vee, b^\vee \in \widetilde{D}^\vee(\Phi)$ which are not orthogonal and such that $n_{\overline{a}} \neq n_{\overline{b}}$. If Φ is simply laced, it follows from Proposition 6.3.2 that $n(\pi(a^\vee), \pi(b^\vee)) \neq n(\pi(b^\vee), \pi(a^\vee))$, and hence that $\Phi^{\mathrm{proj}}(\ell)$ is not simply laced. If Φ is not simply laced, $a^\vee$ is a short coroot, and $b^\vee$ is a long coroot, it is easy to see that $\mathrm{Stab}(a^\vee) \subseteq \mathrm{Stab}(b^\vee)$, and thus that $n_{\overline{a}} \geq n_{\overline{b}}$. Thus $n(\pi(a^\vee), \pi(b^\vee)) = n(a^\vee, b^\vee)$, and

$$n(\pi(b^\vee), \pi(a^\vee)) \geq n(b^\vee, a^\vee) > n(a^\vee, b^\vee).$$

Thus $\Phi^{\mathrm{proj}}(\ell)$ is non-simply laced in this case also.

Direct inspection then handles the case where Φ is of type A and ℓ is not cyclic. □

In the next section we will prove a related result (Proposition 6.3.4) which also covers the remaining case when Φ has a component of type A_n whose stabilizer is a group of rotations.

6.4. The case of an outer automorphism

For future reference, we want to work out the results of Sections 6.1 and 6.2 in the case where $\tau = \ell$. In this case, τ is induced from a group of diagram automorphisms of the Dynkin diagram of Φ. Hence τ acts on the extended diagram, fixing the extended root. The results described here are due, for the most part, to de Siebenthal [**17**]. Notice that if $\tau \neq \mathrm{Id}$ and Φ is irreducible, there are very few possibilities: Φ is simply laced and is of type A, D, E_6 if τ has order 2, and is of type D_4 if τ has order 3 or 6.

LEMMA 6.4.1. *In the above notation, assume that τ is a group of linear transformations of V.*

1. *For every $a \in \Phi$, $\overline{a} = a|V^\tau$ is nonzero.*
2. *There is a one-to-one correspondence between $\Phi^{\mathrm{res}}(\tau)$ and the set of orbits Φ/τ.*
3. *The subset Δ/τ of $\Phi^{\mathrm{res}}(\tau)$ is a set of simple roots.*
4. *The highest root for the irreducible factors of $\Phi^{\mathrm{res}}(\tau)$ corresponding to the above set of simple roots are the images $\overline{d}$, where d is the highest root of an irreducible factor of Φ.*
5. *The root integer for $\Phi^{\mathrm{res}}(\tau)$ corresponding to $\overline{a}$, for $a \in \widetilde{\Delta}$, is the integer $n_{\overline{a}} h_a$.*
6. *$\Phi^{\mathrm{res}}(\tau)$ is reduced if and only if there are no exceptional orbits. In this case $\Phi^{\mathrm{res}}(\tau) = \Phi^\tau$.*
7. *$\Phi^{\mathrm{res}}(\tau)$ is not reduced if and only if Φ has an irreducible factor of type A_{2k} and the stabilizer of this component in τ is non-trivial. In this case,*

Φ^τ is the subsystem of $\Phi^{\mathrm{res}}(\tau)$ consisting of the roots a such that $2a$ is not a root. The set of indivisible roots in $\Phi^{\mathrm{res}}(\tau)$ is also a root system, and Δ/τ is also a set of simple roots for this root system.

PROOF. (1) follows since V^τ contains a regular element.

To see (2), note that, if a and $a' \in \Phi$ are such that $a|V^\tau = a'|V^\tau$, then $W_a \cap V^\tau = W_{a'} \cap V^\tau$. By Lemma 6.2.2, this can only happen if a and $\pm a'$ lie in the same orbit or a lies in an exceptional orbit and $\pm a' = a + \tau(a)$. In this case $a'|V^\tau = \pm 2a|V^\tau$. Thus, if $a|V^\tau = a'|V^\tau$ and a and a' do not lie in the same orbit, then $ka|V^\tau = 0$ for some $k > 0$. This contradicts (1). Moreover, it follows that $\Phi^{\mathrm{res}}(\tau)$ is reduced if and only if there are no exceptional orbits.

It suffices to prove (3), (4), (5), (6), and (7) under the assumption that Φ is irreducible. To see (3), (4), and (5), note that every positive root b can be written as a positive integral linear combination $\sum_{a\in\Delta} r_a a$. Thus $\bar{b} = \sum_{\bar{a}} r_a n_{\bar{a}} \bar{a}$, and similarly for negative roots. Since the cardinality of Δ/τ is the dimension of V^τ, it follows that the $\bar{a}$ are linearly independent and hence a set of simple roots for $\Phi^{\mathrm{res}}(\tau)$. Taking $b = d$, we see that the coefficients of $\bar{b}$, in terms of the simple roots, are at most those of $\bar{d}$, and hence $\bar{d}$ is a highest root for $\Phi^{\mathrm{res}}(\tau)$. Hence the root integers are as claimed. To see (6), since $\bar{d}$ is the highest root for $\Phi^{\mathrm{res}}(\tau)$, it follows that the alcove for Φ^τ is the alcove for $\Phi^{\mathrm{res}}(\tau)$. Thus if $\Phi^{\mathrm{res}}(\tau)$ is reduced we must have $\Phi^{\mathrm{res}}(\tau) = \Phi^\tau$. If $\Phi^{\mathrm{res}}(\tau)$ is not reduced, it is easy to see that the root system associated to the alcove is exactly the set of roots a such that $2a$ is not a root. The remaining statement in (7) follows by a direct inspection. □

Next we determine the Weyl group and the coroot lattice.

LEMMA 6.4.2. *If τ is a group of linear transformations of V,*

(1) *$W(V^\tau, \Phi) = W(\Phi^{\mathrm{proj}}(\tau)) = W(\Phi^\tau) = Z_W(\tau)$, the subgroup of elements of W which commute with τ.*
(2) *The coroot lattice for $\Phi^\tau = \Phi^{\mathrm{res}}(\tau)$ is $(Q^\vee)^\tau = Q^\vee \cap V^\tau$.*
(3) *The coroot lattice for $\Phi^{\mathrm{proj}}(\tau)$ is $\pi(Q^\vee)$.*

PROOF. Clearly $Z_W(\tau)$ normalizes V^τ, and since V^τ contains a regular element, the action of $Z_W(\tau)$ on V^τ is faithful. Thus $Z_W(\tau) \subseteq W(V^\tau, \Phi)$. Conversely, if $w \in W(V^\tau, \Phi)$, choose a regular element $x \in V^\tau$. Then, for all $g \in \tau$, $w(x) = g(w(x)) = g(w(g^{-1}(x)))$. Thus since x is regular $w = g \circ w \circ g^{-1}$ for all $g \in \tau$, so that $w \in Z_W(\tau)$.

To see (2), the coroot lattice for $\Phi^\tau = \Phi^{\mathrm{res}}(\tau)$ is spanned by elements of the form $\epsilon(\bar{a}) \sum_{b\in\bar{a}} b^\vee$. If $\bar{a}$ is ordinary, this is just $\sum_{b\in\bar{a}} b^\vee$, and if $\bar{a}$ is exceptional, corresponding to the pair $\prod_i \{a_{i,1}, a_{i,2}\}$, then $a_{i,1}^\vee + a_{i,2}^\vee$ is again in the coroot lattice and the corresponding orbit sum is $\sum_i a_{i,1}^\vee + a_{i,2}^\vee$. Thus the coroot lattice is generated by orbit sums. Since τ permutes an integral basis for $Q^\vee$, the coroot orbits generate $(Q^\vee)^\tau$.

Finally, it follows from the definition that the coroot lattice for $\Phi^{\mathrm{proj}}(\tau)$ is $\pi(Q^\vee)$. □

Using Proposition 6.3.2 and Lemma 6.2.7, we see that, in case Φ is irreducible and τ is nontrivial, $\Phi^{\mathrm{res}}(\tau)$ is given as follows:

- If Φ is of type A_{2n-2}, $n > 1$, then $\Phi^{\mathrm{res}}(\tau)$ is of type BC_n.
- If Φ is of type A_{2n-1}, $n > 1$, then $\Phi^{\mathrm{res}}(\tau)$ is of type C_n.
- If Φ is of type D_{n+1}, $n \geq 3$, and τ has order 2 then $\Phi^{\mathrm{res}}(\tau)$ is of type B_n.

- If Φ is of type D_4, and τ has order 3, then $\Phi^{\rm res}(\tau)$ is of type G_2.
- If Φ is of type E_6, then $\Phi^{\rm res}(\tau)$ is of type F_4.

6.5. Further results under an additional hypothesis

We now make one further assumption on the group τ:

ASSUMPTION 6.5.1. *The fixed subspace V^ℓ of the linearization ℓ of τ is written as the intersection of the kernels of a subset of the roots of Φ.*

Let $\Phi^\perp$ be the subset of Φ consisting of roots vanishing on V^ℓ. Let U be the subspace spanned by the coroots inverse to the roots in $\Phi^\perp$. Then $\Phi^\perp$ is a root system on U. The above assumption on V^ℓ implies that $U \oplus V^\ell = V$. Let $\pi\colon V \to V^\ell$ denote orthogonal projection.

LEMMA 6.5.2. *The intersection $V^\tau \cap U$ is the barycenter of an alcove for $\Phi^\perp$.*

PROOF. Let $\tilde{x}_0 = V^\tau \cap U$. We claim that no root of $\Phi^\perp$ is integral on $\tilde{x}_0$ and hence $\tilde{x}_0$ is in the interior of an alcove $B \subset U$ of the root system $\Phi^\perp$. In fact, if $a \in \Phi^\perp$ is integral on $\tilde{x}_0$, then a is a root of Φ which vanishes on V^ℓ and hence is integral on $\{\tilde{x}_0\} + V^\ell = V^\tau$. But this contradicts our assumption that V^τ contains an interior point of an alcove for Φ. The group τ normalizes U and the alcove structure for the root system $\Phi^\perp$. The point $\tilde{x}_0$ is the unique fixed point for $\tau|U$ and is in the interior of B. Thus $\tilde{x}_0$ is the barycenter of B. □

PROPOSITION 6.5.3. *The Weyl group $W(V^\ell, \Phi)$ is identified with the Weyl group of Φ^τ.*

PROOF. Let $g \in Z_{W_{\rm aff}(\Phi)}(\tau)$. Then its differential is an element of the Weyl group of Φ which normalizes V^ℓ. The elements of the Weyl group of Φ^τ are exactly the restrictions to V^ℓ of the differentials of elements $g \in Z_{W_{\rm aff}(\Phi)}(\tau)$. This proves that the Weyl group of Φ^τ is identified with a subgroup of $W(V^\ell, \Phi)$.

Let $g \in N_{W(\Phi)}(V^\ell)$. Then g normalizes U and the alcove decomposition of U for $\Phi^\perp$. As such, setting $\tilde{x}_0 = V^\tau \cap U$, the point $g \cdot \tilde{x}_0$ is the barycenter of some alcove for this alcove decomposition. Thus, there is an element h of the affine Weyl group $W_{\rm aff}(\Phi^\perp)$ with $h \cdot \tilde{x}_0 = g \cdot \tilde{x}_0$. Then $h^{-1}g(\tilde{x}_0) = \tilde{x}_0$. Let $w \in W(\Phi^\perp)$ be the Weyl part of h. Since $w|V^\ell = \mathrm{Id}$, the element $w^{-1}g$ normalizes V^ℓ. Hence, $h^{-1}g$ normalizes $\{\tilde{x}_0\} + V^\ell = V^\tau$, and hence by Lemma 6.1.3 is an element of $Z_{W_{\rm aff}(\Phi)}(\tau)$. Thus, the restriction to V^ℓ of its differential is an element of the Weyl group of Φ^τ. The element h, and consequently also its differential, centralize V^ℓ. Thus, the restriction to V^ℓ of the differential of $h^{-1}g$ agrees with that of g. This proves that the Weyl group of Φ^τ is all of $W(V^\ell, \Phi)$. □

REMARK 6.5.4. Unlike the case of a group of linear automorphisms, it is not in general true that $W(V^\ell, \Phi)$ is equal to $Z_W(\ell)$, the set of $w \in W$ which commute with ℓ. However, there is a surjection from $Z_W(\ell)$ to $W(V^\ell, \Phi)$, and in fact $Z_W(\ell) \cong I \rtimes W(V^\ell, \Phi)$ for an appropriate subgroup I of $Z_W(\ell)$.

Proposition 6.5.3 gives an extension of Proposition 6.3.4 to the case of an arbitrary group τ:

PROPOSITION 6.5.5. *Suppose that τ is a group of affine isometries of V normalizing an alcove of Φ. Then the Weyl group of $W(V^\ell, \Phi)$ is identified with the Weyl group of $\Phi^{\rm res}(\ell)$ or equivalently with the Weyl group of $\Phi^{\rm proj}(\ell)$.*

PROOF. Clearly, we may assume that Φ is irreducible. The only irreducible case not covered by Proposition 6.3.4 is when Φ is of type A and ℓ is a group of rotations of the extended Dynkin diagram. We choose a representation of Φ on the subspace $\{(x_1, \ldots, x_{n+1}) | \sum x_i = 0\}$ in $\mathbf{R}^{n+1}$ such that the set of simple roots is $\{e_i - e_{i+1}\}$. Suppose that ℓ is a group of order m and let $k = (n+1)/m$. Then V^ℓ is the intersection of the kernels of the roots $e_i - e_{i+k}$. Hence, this case is covered by Proposition 6.5.3. □

Next we describe the coroot lattice of Φ^τ.

PROPOSITION 6.5.6. *The coroot lattice of the affine Weyl group associated to Φ^τ is identified with $\pi(Q^\vee(\Phi))$.*

PROOF. We decompose $V = U \oplus V^\ell$ and let $\tilde{x}_0 = U \cap V^\tau$. Suppose that $g \in Z_{W_{\text{aff}}(\Phi)}(\tau)$ and that $g|V^\tau$ is translation by an element $\gamma_2 \in V^\ell$. We write

$$g(a_1, a_2) = (wa_1 + \gamma_1, b + \gamma_2)$$

for w in the Weyl group of $\Phi^\perp$. Since $g \in W_{\text{aff}}(\Phi)$, $\gamma_1 + \gamma_2 \in Q^\vee(\Phi)$, and hence $\gamma_2 \in \pi(Q^\vee(\Phi))$.

Conversely, if $\gamma_2 \in \pi(Q^\vee(\Phi)$, write $\gamma = \gamma_1 + \gamma_2$ for some $\gamma \in Q^\vee(\Phi)$, with $\gamma_1 \in U$. Clearly, γ_1 is contained in the center of the root system $\Phi^\perp$, and hence $\tilde{x}_0 + \gamma_1$ is the barycenter of an alcove for $\Phi^\perp$. This means that there is an element $h \in W_{\text{aff}}(\Phi^\perp)$ such that $h(\tilde{x}_0 + \gamma_1) = \tilde{x}_0$. The composition of translation by γ followed by h is an element of $Z_{W_{\text{aff}}(\Phi)}(\tau)$ whose restriction to V^τ is translation by γ_2. □

6.6. The case of a subgroup of $\mathcal{C}\Phi$

LEMMA 6.6.1. *Let $\mathcal{C} \subseteq \mathcal{C}\Phi$ be a subgroup, and let τ be the group of affine automorphisms of V normalizing A determined by the action of $\mathcal{C}$ on the alcove. Then τ satisfies the hypothesis of Assumption 6.5.1.*

PROOF. Since Φ is reduced, we may assume that $\Phi = \Phi(T, G)$ for some G and that $V = i\mathfrak{t}$. By Proposition 3.5.4, for each $c \in \mathcal{C}$, the fixed point subspace $\mathfrak{t}^{w_c}$ is conjugate under the Weyl group to $\mathfrak{t}_c$. Since $i\mathfrak{t}_c$ satisfies the hypothesis of Assumption 6.5.1, so does $i\mathfrak{t}^{w_c} = V^{w_c}$. Since $w_{\mathcal{C}}$ is the linearization of τ, and $\mathfrak{t}^{w_{\mathcal{C}}} = \bigcap_{c \in \mathcal{C}} \mathfrak{t}^{w_c}$, the lemma follows. □

PROPOSITION 6.6.2. *The root system Φ^τ is identified with the subroot system of $\Phi^{\text{proj}}(w_{\mathcal{C}})$ consisting of all roots whose inverse coroots are indivisible elements of $\Phi^{\text{proj}}(w_{\mathcal{C}})^\vee$.*

PROOF. By Proposition 6.5.6 and Proposition 6.2.6, Φ^τ and $\Phi^{\text{proj}}(w_{\mathcal{C}})$ have the same coroot lattices. By Proposition 6.2.6, their Weyl groups are the same. Now there is the following general lemma on two root systems with the same coroot lattice:

LEMMA 6.6.3. *Let Φ_1 and Φ_2 be two root systems on a vector space V, and suppose that Φ_1 is reduced. Suppose that the coroot lattice $Q^\vee(\Phi_1)$ is equal to the coroot lattice $Q^\vee(\Phi_2)$ and that $W(\Phi_1) = W(\Phi_2)$. Then Φ_1 is the set of all non-multipliable roots in Φ_2. In particular, if Φ_2 is reduced, then $\Phi_1 = \Phi_2$.*

PROOF. Let Φ_2' be the subroot system of Φ_2 consisting of the non-multipliable roots. Then $(\Phi_2')^\vee$ is the sub-coroot system of $\Phi_2^\vee$ consisting of all indivisible coroots in $\Phi_2^\vee$. In particular, the coroot lattice of Φ_2' is equal to that of Φ_2. Of course, Φ_2 and Φ_2' have the same Weyl group. Thus, it suffices to assume that Φ_2 is reduced.

Let $Q^\vee = Q^\vee(\Phi_1) = Q^\vee(\Phi_2)$. For each $a \in \Phi_1$, $a^\vee$ is an indivisible element of $Q^\vee$. Since reflection in the wall defined by a is an element of $W(\Phi_1) = W(\Phi_2)$, there exists a $b \in \Phi_2$ such that $a^\vee = rb^\vee$ for some real number r. But $b^\vee \in Q^\vee$ is also indivisible, so that $r = \pm 1$. Thus $\Phi_1^\vee \subseteq \Phi_2^\vee$. By symmetry $\Phi_2^\vee \subseteq \Phi_1^\vee$, and hence $\Phi_1^\vee = \Phi_2^\vee$. □

Applying the lemma to $\Phi_1 = \Phi^\tau$ and $\Phi_2 = \Phi^{\mathrm{proj}}(w_C)$ completes the proof of Proposition 6.6.2. □

PROPOSITION 6.6.4. *Suppose that Φ is irreducible. Orthogonal projection induces an embedding $\widetilde{\Delta}_C^\vee$ in V^{w_C}. Its image is a set of coroots inverse to an extended set of simple roots either for $\Phi^{\mathrm{proj}}(w_C)$ or for the subroot system consisting of all non-multipliable roots of $\Phi^{\mathrm{proj}}(w_C)$.*

PROOF. We have an identification of Φ^τ with the non-multipliable roots of $\Phi^{\mathrm{proj}}(w_C)$. We also know by Proposition 6.1.5 that, up to positive multiples, the image under orthogonal projection of $\widetilde{\Delta}_C^\vee$ forms the set of coroots inverse to an extended set of simple coroots for Φ^τ. If $\Phi^{\mathrm{proj}}(w_C)$ is reduced, then it is equal to Φ^τ and its coroots are indivisible elements in the coroot lattice. Thus, as before, the multiples are all $+1$ and the image of $\widetilde{\Delta}_C^\vee$ under orthogonal projection forms the set of coroots inverse to an extended set of simple roots for $\Phi^{\mathrm{proj}}(w_C)$.

If $\Phi^{\mathrm{proj}}(w_C)$ is not reduced, then it is of type BC_n for some $n \geq 1$ and Φ^τ is the subsystem of type C_n. In the extended set of simple coroots for Φ^τ all but two are neither divisible nor multipliable in $\Phi^{\mathrm{proj}}(w_C)$ and the last two are multipliable by 2 in $\Phi^{\mathrm{proj}}(w_C)$. Thus, the image of $\widetilde{\Delta}_C^\vee$ contains all the former coroots and, for each of the latter two, contains either the coroot in Φ^τ or twice it. A priori there are three possibilities: (i) $\widetilde{\Delta}_C^\vee$ is equal to the extended set of coroots for Φ^τ (i.e. consists of indivisible coroots); (ii) $\widetilde{\Delta}_C^\vee$ is equal to the extended set of coroots for $\Phi^{\mathrm{proj}}(w_C)$ (i.e., contains one indivisible but multipliable coroot and one non-multipliable but divisible coroot); or (iii) contains two non-multipliable but divisible coroots. In case (iii) the lattice spanned by $\widetilde{\Delta}_C^\vee$ is of index two in the coroot lattice of $\Phi^{\mathrm{proj}}(w_C)$ and hence this case is ruled out by Proposition 6.5.6. Since by Proposition 6.6.2 Φ^τ is the subroot system of $\Phi^{\mathrm{proj}}(w_C)$ consisting of the non-multipliable roots, cases (i) and (ii) are exactly the two cases listed in the statement of the proposition. □

DEFINITION 6.6.5. We define $\Phi(w_C)$ to be the subroot system of $\Phi^{\mathrm{proj}}(w_C)$ such that the image of $\widetilde{\Delta}_C^\vee$ in V^{w_C} is an extended set of simple coroots. By the previous proposition, $\Phi(w_C)$ is either $\Phi^{\mathrm{proj}}(w_C)$ or the subroot system consisting of all non-multipliable roots of $\Phi^{\mathrm{proj}}(w_C)$.

COROLLARY 6.6.6. *There is a positive integer n_0 such that, for all $\overline{a} \in \widetilde{\Delta}_C$, $g_{\overline{a}}$ is equal to n_0 times the coroot integer $m_{\overline{a}}$ for the extended simple root of $\Phi(w_C)$ corresponding to $\overline{a}$.*

PROOF. By Equation 6.2 in Section 6.3, $\sum_{\overline{a}\in\widetilde{\Delta}_C} g_{\overline{a}}\pi(a^\vee) = 0$. On the other hand, by definition $\sum_{\overline{a}\in\widetilde{\Delta}_C} m_{\overline{a}}\pi(a^\vee) = 0$, and the $m_{\overline{a}}$ are relatively prime integers. Thus the corollary is clear. □

6.7. Proof of Theorem 1.6.2

Let $\mathcal{C}$ be a subgroup of $\mathcal{C}G$ and let τ be the corresponding group of affine isometries. By Lemma 6.6.1, τ satisfies Assumption 6.5.1. Thus, the results of this section apply to τ. By Proposition 6.6.4, π embeds $\widetilde{\Delta}_{\mathcal{C}}^{\vee}$ as an extended set of simple coroots for either the root system $\Phi^{\mathrm{proj}}(w_{\mathcal{C}})$ or for the subroot system of nonmultipliable roots in $\Phi^{\mathrm{proj}}(w_{\mathcal{C}})$. In either case, we let $\Phi(w_{\mathcal{C}})$ be the corresponding root system. The Weyl group of $\Phi(w_{\mathcal{C}})$ is the same as the Weyl group of $\Phi^{\mathrm{proj}}(w_{\mathcal{C}})$, and by Proposition 6.5.3, this is $W(S^{w_{\mathcal{C}}}, G)$. The coroot lattice of $\Phi(w_{\mathcal{C}})$ is the lattice generated by the indivisible coroots in $\Phi^{\mathrm{proj}}(w_{\mathcal{C}})^{\vee}$, which is is the coroot lattice of $\Phi^{\mathrm{proj}}(w_{\mathcal{C}})$. By Proposition 6.2.6, the coroot lattice of $\Phi^{\mathrm{proj}}(w_{\mathcal{C}})$ is $\pi(Q^{\vee})$. By Proposition 6.3.2, the Cartan matrix associated to $\widetilde{\Delta}_{\mathcal{C}}^{\vee}$ agrees with the Cartan matrix associated to $\widetilde{D}_{\mathcal{C}}^{\vee}$ as given in Definition 1.6.1. This completes the proof of Theorem 1.6.2.

CHAPTER 7

The fixed subgroup of an automorphism

Our main goal in this chapter is to describe the component group of the centralizer of a c-pair in G. In order to do so, it will be convenient to study more generally the fixed subgroups of certain automorphisms of compact, connected groups. In particular, we shall see how the fixed subgroup changes as we vary the given automorphism by composing with an inner automorphism. The fixed subgroup is completely described by its Lie algebra, its component group, and the fundamental group of the identity component. We shall give a complete description of the component group of the fixed subgroup, and indicate briefly how the methods of this paper can be used to identify the Lie algebra. It is also straightforward to describe the fundamental group of its identity component, or of the derived subgroup of the identity component, but we shall not do this here.

In this chapter, H is a compact, connected group with Lie algebra $\mathfrak{h}$ and σ is an automorphism of H whose restriction to the center of H has finite order. The fixed set of σ will be denoted by H^σ. For $h \in H$, we let $\sigma_h = i_h \circ \sigma$. It is an easy exercise to check that, if $h = g^{-1}\sigma(g)$ for some $g \in H$, then H^{σ_h} is conjugate to H^σ in H.

7.1. A first description of the component group

LEMMA 7.1.1. *There exists a maximal torus T of H invariant under σ such that $(T^\sigma)^0$ is a maximal torus of H^σ. For such a maximal torus T, we set $S^\sigma = (T^\sigma)^0$. Then*

$$(N_H(T))^\sigma = N_{H^\sigma}(S^\sigma).$$

PROOF. By **[17]**, applied to the derived subgroup DH of H, there exists an element X in $\mathfrak{h}^\sigma$ which is regular in $\mathfrak{h}$, and hence *a fortiori* in $\mathfrak{h}^\sigma$. Let T be the unique maximal torus in H whose Lie algebra contains X, and let S^σ be the unique maximal torus in H^σ whose Lie algebra contains X. Clearly $S^\sigma = (T^\sigma)^0$ and $(N_H(T))^\sigma = N_{H^\sigma}(S^\sigma)$. □

We assume that T satisfies the conclusions of the the previous lemma, and let $X \in \mathfrak{h}^\sigma$ be a regular element in $\mathfrak{h}$. Let $\Phi_H = \Phi(T, H)$, let Φ_H^+ be the set of positive roots defined by

$$\Phi_H^+ = \{a \in \Phi_H | a((2\pi i)^{-1}X) > 0\},$$

and let Δ_H be the corresponding set of simple roots. The automorphism $\sigma\colon \mathfrak{t} \to \mathfrak{t}$ extends to $\mathfrak{t} \otimes_{\mathbf{R}} \mathbf{C}$ and induces an automorphism $\sigma\colon i\mathfrak{t} \to i\mathfrak{t}$ which normalizes Φ_H and Δ_H. Thus, σ normalizes the alcove $A \subset i\mathfrak{t}$ corresponding to Δ_H. The σ-orbit of $a \in \Delta_H$ is denoted $\overline{a}$. All elements of $\overline{a}$ have the same restriction to $i\mathfrak{t}^\sigma$. Thus, we can view $\overline{a}$ as an element of $(i\mathfrak{t}^\sigma_{\mathbf{C}})^*$. For each $\overline{a} \in \Delta_H/\langle\sigma\rangle$ we denote by $n_{\overline{a}}$ the cardinality of the orbit $\overline{a} \subseteq \Delta_H$.

DEFINITION 7.1.2. An automorphism σ which normalizes T and an alcove $A \subset i\mathfrak{t}$ containing the origin is said to be *in normal form* with respect to T and A.

LEMMA 7.1.3. *Fix a maximal torus T of H and an alcove $A \subset i\mathfrak{t}$ containing the origin. Let σ be an automorphism of H. Then there exists an $h \in H$ of the form $g^{-1}\sigma(g)$ such that σ_h is in normal form with respect to T and A. Moreover, H^σ and H^{σ_h} are conjugate in H.*

PROOF. By the previous lemma and remarks, there exists some maximal torus T' and an alcove $A' \subset i\mathfrak{t}$ containing the origin such that σ normalizes T' and A'. Let $g \in H$ be an element such that $gT'g^{-1} = T$ and $gA'g^{-1} = A$. Then $h = g^{-1}\sigma(g)$ satisfies the conclusions of the lemma. □

We assume that σ is in normal form with respect to T and A. Then σ acts on $\mathfrak{t}$, as does $W = W(T,H)$. Let $Z_W(\sigma)$ be the subgroup of elements of W which commute with σ (all viewed as linear automorphisms of $\mathfrak{t}$). The inclusion $(N_H(T))^\sigma \subseteq N_H(T)$, followed by the projection from $N_H(T)$ to W, induces a homomorphism from $(N_H(T))^\sigma$ to W. Its kernel is clearly T^σ. Denote by $N(\sigma)$ the image of $(N_H(T))^\sigma$ in W. Clearly $N(\sigma) \subseteq Z_W(\sigma)$, and there is an exact sequence

$$\{1\} \to T^\sigma \to (N_H(T))^\sigma \to N(\sigma) \to \{1\}.$$

LEMMA 7.1.4. *Every component of H^σ contains an element normalizing $\mathfrak{t}^\sigma$. If g and g' are in the same component of H^σ and normalize $\mathfrak{t}^\sigma$, then there exists an element $w \in W(S^\sigma, (H^\sigma)^0)$ such that $g'|\mathfrak{t}^\sigma = w \circ (g|\mathfrak{t}^\sigma)$.*

PROOF. Let $g \in H^\sigma$. Then $gS^\sigma g^{-1}$ is a maximal torus of H^σ and hence there is $h \in (H^\sigma)^0$ such that $(gh)S^\sigma(gh)^{-1} = S^\sigma$. Clearly, $gh \in N_H(S^\sigma)$ and gh and g lie in the same component of H^σ. If g and g' are in the same component of H^σ and both normalize $\mathfrak{t}^\sigma$, then $g'g^{-1} \in (H^\sigma)^0$ and normalizes $\mathfrak{t}^\sigma$. Thus $g'g^{-1}|\mathfrak{t}^\sigma = w$ for some $w \in W(S^\sigma, (H^\sigma)^0)$. □

LEMMA 7.1.5. *The inclusion of $N_H(T)^\sigma$ in H^σ induces a homomorphism from $N_H(T)^\sigma$ to $\pi_0(H^\sigma)$ whose kernel is $N_{(H^\sigma)^0}(S^\sigma)$. Thus there is an isomorphism*

$$N_H(T)^\sigma / N_{(H^\sigma)^0}(S^\sigma) \cong \pi_0(H^\sigma).$$

PROOF. The induced map $N_H(T)^\sigma$ to $\pi_0(H^\sigma)$ is surjective by Lemma 7.1.4. Its kernel is $N_H(T)^\sigma \cap (H^\sigma)^0$. Any element in this intersection normalizes T and its action on T commutes with σ. Thus, it also normalizes S^σ. Conversely, since S^σ contains a regular element of T, any element of $(H^\sigma)^0$ normalizing S^σ normalizes T. □

LEMMA 7.1.6. *There is an exact sequence*

$$\{1\} \to \pi_0(T^\sigma) \to N_H(T)^\sigma / N_{(H^\sigma)^0}(S^\sigma) \to N(\sigma)/W(S^\sigma, (H^\sigma)^0) \to \{1\}.$$

PROOF. By definition, there is an exact sequence

$$\{1\} \to T^\sigma \to N_H(T)^\sigma \to N(\sigma) \to \{1\}.$$

Now $N_{(H^\sigma)^0}(S^\sigma)$ is a normal subgroup of $N_H(T)^\sigma$, and the image of $N_{(H^\sigma)^0}(S^\sigma)$ in $W(T,H)$ is $W(S^\sigma, (H^\sigma)^0)$. Thus $N(\sigma)$ contains $W(S^\sigma, (H^\sigma)^0)$ as a normal subgroup, and there is a surjection

$$N_H(T)^\sigma / N_{(H^\sigma)^0}(S^\sigma) \to N(\sigma)/W(S^\sigma, (H^\sigma)^0).$$

By a diagram chase, the kernel of this map is then

$$T^\sigma/N_{(H^\sigma)^0}(S^\sigma)\cap T = T^\sigma/S^\sigma = \pi_0(T^\sigma).$$

This establishes the exact sequence of the statement. □

COROLLARY 7.1.7. *There is an exact sequence*

$$\{1\}\to\pi_0(T^\sigma)\to\pi_0(H^\sigma)\to N(\sigma)/W(S^\sigma,(H^\sigma)^0)\to\{1\}.$$

In the case of an inner automorphism, this result is due to Steinberg [**19**].

LEMMA 7.1.8. *Let $\mathfrak{s}^\sigma$ be the Lie algebra of S^σ and let $X\in\mathfrak{h}^\sigma$ be a regular element contained in $\mathfrak{t}$. Let C_0 be the unique chamber in $i\mathfrak{s}^\sigma$ for the group $W(S^\sigma,(H^\sigma)^0)$ which contains $(2\pi i)^{-1}X$. If C_0 is contained in a Weyl chamber C of $i\mathfrak{t}$, then $N(\sigma)=W(S^\sigma,(H^\sigma)^0)$ and hence $\pi_0(H^\sigma)=\pi_0(T^\sigma)$.*

PROOF. Let $w\in N(\sigma)$, and lift w to an element g of $N_H(T)$ commuting with σ. Then $w((2\pi i)^{-1}X) = \mathrm{Ad}(g)((2\pi i)^{-1}X)$ lies in a unique Weyl chamber C_1 of $i\mathfrak{s}^\sigma$. After multiplying by an element of $W(S^\sigma,(H^\sigma)^0)$, we can assume that $C_1=C_0$. But then C and $w(C)$ are two Weyl chambers in $i\mathfrak{t}$ containing C_0 and hence $(2\pi i)^{-1}X$, so that $C=w(C)$ and $w=\mathrm{Id}$. Thus $N(\sigma)=W(S^\sigma,(H^\sigma)^0)$. □

The hypothesis of the lemma does not always hold. In the following section, we will give a condition on σ which will imply that C_0 is contained in exactly one Weyl chamber C in $i\mathfrak{t}$ for $W(H)$.

7.2. Special automorphisms

We keep the notation of the previous section and assume that $\sigma\colon H\to H$ is in normal form with respect to T and A, and that $S^\sigma=(T^\sigma)^0$, with Lie algebra $\mathfrak{s}^\sigma$.

LEMMA 7.2.1. *The set of nonzero weights of the action of $\mathfrak{s}^\sigma$ on $\mathfrak{h}\otimes_{\mathbf{R}}\mathbf{C}$ is identified with the restricted root system $\Phi_H^{\mathrm{res}}(\sigma)$, or equivalently with $\Phi_H/\langle\sigma\rangle$. The set of roots of H^σ with respect to S^σ is a sub-root system of $\Phi_H^{\mathrm{res}}(\sigma)$.*

PROOF. The action of $\mathfrak{s}^\sigma$ on $\mathfrak{h}\otimes_{\mathbf{R}}\mathbf{C}$ normalizes the root spaces of $\mathfrak{h}\otimes_{\mathbf{R}}\mathbf{C}$. The action of $\mathfrak{s}^\sigma$ on a root space $\mathfrak{h}^a$ is given by the weight $a|\mathfrak{s}^\sigma$. Since $\mathfrak{s}^\sigma$ contains a regular element of $\mathfrak{h}$, none of these weights is zero. Hence the weight space of $\mathfrak{h}\otimes_{\mathbf{R}}\mathbf{C}$ corresponding to the weight zero is $\mathfrak{t}\otimes_{\mathbf{R}}\mathbf{C}$. By Lemma 6.4.1, the nonzero weights are identified with $\Phi_H^{\mathrm{res}}(\sigma)$, or equivalently with $\Phi_H/\langle\sigma\rangle$. The roots for the $\mathfrak{s}^\sigma$-action on $\mathrm{Lie}(H^\sigma)\otimes_{\mathbf{R}}\mathbf{C}$ are given by restrictions of roots of Φ_H to S^σ and so are elements of $\Phi_H^{\mathrm{res}}(\sigma)$. Hence the roots for H^σ with respect to S^σ form a subset of $\Phi_H^{\mathrm{res}}(\sigma)$. To see that they are a sub-root system, in other words that the corresponding coroots are the same, it is enough to show that the Weyl invariant inner product on $\mathfrak{t}$ is invariant under the Weyl group of $(H^\sigma)^0$. Since $\mathfrak{s}^\sigma$ contains a regular element of $\mathfrak{t}$, every element in $W(S^\sigma,(H^\sigma)^0)$ is the restriction of an element of $W(T,H)$, and thus the last statement is clear. □

If $s\in S^\sigma$, then $S^{\sigma_s}=S^\sigma$ and $T^{\sigma_s}=T^\sigma$. However, H^{σ_s} and H^σ need not be conjugate in H. In fact, H^{σ_s} and H^σ need not have the same dimension or the same number of connected components.

LEMMA 7.2.2. *There is an element $s\in S^\sigma$ such that the set $\Delta_H/\langle\sigma\rangle$ is a set of simple roots for H^{σ_s} with respect to the maximal torus $S^\sigma=S^{\sigma_s}$.*

PROOF. Since Δ_H is a linearly independent subset of $\mathfrak{t}^*$ invariant under σ, the quotient set $\Delta_H/\langle\sigma\rangle$ is a linearly independent subset of $(\mathfrak{t}^\sigma)^*$. For each $\overline{a} \in \Delta_H^\vee/\langle\sigma\rangle$, order $\overline{a} = \{a_1, \ldots, a_{n_{\overline{a}}}\}$ so that $\sigma{\cdot}a_i = a_{i+1}, 1 \leq i \leq n_{\overline{a}}$ (by convention $a_{n_{\overline{a}}+1} = a_1$). Then the action of σ on $\mathfrak{h} \otimes_{\mathbf{R}} \mathbf{C}$ sends the root space $\mathfrak{h}^{a_i}$ to $\mathfrak{h}^{a_{i+1}}$. Hence, $\sigma^{n_{\overline{a}}}$ sends $\mathfrak{h}^{a_i}$ to itself for every i. This means that there is $q_{a_i} \in U(1)$ such that $\sigma^{n_{\overline{a}}}|\mathfrak{h}^{a_i}$ is multiplication by q_{a_i} on each root space $\mathfrak{h}^{a_i}$. By equivariance under σ, we see that $q_{a_i} = q_{a_j}$ for all $1 \leq i, j \leq n_{\overline{a}}$. We denote this common value by $q_{\overline{a}}$. Choose an element $\mu_{\overline{a}} \in U(1)$ such that $\mu_{\overline{a}}^{n_{\overline{a}}} = q_{\overline{a}}$. Since the elements of $\Delta_H/\langle\sigma\rangle$ are linearly independent in $(\mathfrak{t}^\sigma)^*$, it follows that there is an element $s \in S^\sigma$ such that $s^a = \mu_{\overline{a}}^{-1}$ for every $a \in I$.

By construction, if $\overline{a} = \{a_1, \ldots, a_{n_{\overline{a}}}\}$, then $\sigma_s^{n_{\overline{a}}}|\bigoplus_i \mathfrak{h}^{a_i}$ is the identity. Choose a non-zero element $X_a \in \mathfrak{h}^a$. Set

$$X_{\overline{a}} = X_a + \sigma_s X_a + \cdots + \sigma_s^{n_{\overline{a}}-1} X_a.$$

Then $X_{\overline{a}}$ is a non-zero element of $\mathfrak{h} \otimes_{\mathbf{R}} \mathbf{C}$ invariant under σ_s, and hence $X_{\overline{a}} \in \mathrm{Lie}(H^{\sigma_s}) \otimes_{\mathbf{R}} \mathbf{C}$. Moreover, for all $s \in S^\sigma$, $\mathrm{Ad}(s)X_{\overline{a}} = \overline{a}(s)X_{\overline{a}}$. Thus, each $\overline{a} \in \Delta_H/\langle\sigma\rangle$ is a root of H^{σ_s} with respect to S^σ. Since every root of H^{σ_s} is the restriction of a root of H to S^σ and since Δ_H is a set of simple roots for H, it follows that every root of H^{σ_s} can be written uniquely as a linear combination of the elements in $\Delta_H/\langle\sigma\rangle$ and the coefficients of this linear combination are either all positive integers or all negative integers. Consequently, $\Delta_H/\langle\sigma\rangle$ is a set of simple roots of H^{σ_s} with respect to S^σ. □

DEFINITION 7.2.3. An automorphism σ_s which satisfies the conclusions of the previous lemma is called a *special automorphism*.

LEMMA 7.2.4. *Suppose that σ_s is special. Then every nonzero weight for the action of $\mathfrak{s}^\sigma$ on $\mathfrak{h} \otimes_{\mathbf{R}} \mathbf{C}$ is an integral linear combination of the elements in $\Delta_{H^{\sigma_s}}$ such that all coefficients have the same sign. Moreover $\pi_0(T^{\sigma_s}) = \pi_0(H^{\sigma_s})$.*

PROOF. The first assertion follows from the fact that every element of $\Delta_{H^{\sigma_s}}$ is the restriction of an element of Δ_H. It follows that the Weyl chamber in $\mathfrak{s}^\sigma$ defined by $\Delta_{H^{\sigma_s}}$ is contained in the Weyl chamber defined by Δ_H. By Lemma 7.1.8, it follows that $\pi_0(T^{\sigma_s}) = \pi_0(H^{\sigma_s})$. □

PROPOSITION 7.2.5. *Let σ_s be a special automorphism. Then the roots of H^{σ_s} are the indivisible roots in $\Phi_H^{\mathrm{res}}(\sigma_s)$. In particular, the Weyl group $W(S^{\sigma_s}, H^{\sigma_s})$ is equal to the Weyl group of $\Phi_H^{\mathrm{res}}(\sigma_s)$, namely $Z_W(\sigma_s)$.*

PROOF. According to Lemma 6.4.1, the indivisible roots of $\Phi_H^{\mathrm{res}}(\sigma_s)$ form a reduced root system with $\Delta_H/\langle\sigma_s\rangle$ a set of simple roots. By Lemma 7.2.2, $\Delta_H/\langle\sigma_s\rangle$ is also a set of simple roots for $\Phi(S^{\sigma_s}, (H^{\sigma_s})^0)$. Since the latter is also a reduced root system, the two root systems agree. By Lemma 6.4.2 and Proposition 6.3.4, $W(S^{\sigma_s}, (H^{\sigma_s})^0)$ is then $Z_W(\sigma_s)$. □

The following relates the roots of H^{σ_s}, which are restricted roots, to the projection root system.

COROLLARY 7.2.6. *For σ_s special, there is a natural one-to-one correspondence between the roots of H^{σ_s} and the subroot system of the non-multipliable roots in $\Phi_H^{\mathrm{proj}}(\sigma_s)$. The corresponding roots in these two systems are positive multiples of each other.*

PROOF. According to Proposition 6.5.5, these root systems have the same Weyl group, which implies that they have the same set of walls. Since each of these root systems is reduced, the result follows. □

For example, if Φ_H is simply laced, which is always the case if σ_s is nontrivial and Φ_H is irreducible, and if there are no exceptional orbits, then $H^{\sigma_s} = \Phi_H^{\mathrm{res}}(\sigma_s)$ and $\Phi_H^{\mathrm{proj}}(\sigma_s)$ is the inverse system, and the bijection of the corollary simply associates to each element of H^{σ_s} its inverse coroot.

7.3. A complete description of the component group

Let σ be an automorphism in normal form. We now use the results developed above to give an explicit description of the component group of the fixed subgroup H^σ.

We begin with some general remarks. Let J be an abelian group, let $\mathrm{Tor}\, J$ be the torsion subgroup of J, and let τ be an automorphism of J. Denote by J^τ the subgroup of invariants and by J_τ be the group of coinvariants: $J_\tau = J/\mathrm{Im}(\mathrm{Id} - \tau)J$. Suppose that $\mathfrak{t}$ is a vector space with a τ-invariant inner product, and that Λ is a sublattice of $\mathfrak{t}$ such that $\tau(\Lambda) = \Lambda$. The inner product on $\mathfrak{t}$ identifies $\mathfrak{t}_\tau$ with $\mathfrak{t}^\tau$ via orthogonal projection $\pi\colon \mathfrak{t} \to \mathfrak{t}^\tau$. The projection π induces a map from Λ to $\mathfrak{t}^\tau$ and hence factors through Λ_τ. By comparing ranks, one checks that this sets up an isomorphism from $\Lambda_\tau/\mathrm{Tor}(\Lambda_\tau)$ to $\pi(\Lambda)$. In particular, suppose that τ acts as a permutation representation on a $\mathbf{Z}$-basis of Λ. In this case $\mathrm{Tor}(\Lambda_\tau) = 0$ and π induces an isomorphism from Λ_τ to $\pi(\Lambda)$.

If

$$\{0\} \to J' \to J \to J'' \to \{0\}$$

is an exact sequence of abelian groups such that τ acts on all of the groups and the homomorphisms are τ-equivariant, then there is an associated long exact sequence

$$0 \to (J')^\tau \to J^\tau \to (J'')^\tau \to (J')_\tau \to J_\tau \to (J'')_\tau \to 0.$$

(This is just the long exact group cohomology sequence for the group $\mathbf{Z}$ acting via τ.) Here the homomorphism $\delta\colon (J'')^\tau \to (J')_\tau$ is defined as follows: given $\xi \in (J'')^\tau$, lift ξ to an element $\tilde{\xi} \in J$. Since $\tau(\tilde{\xi}) - \xi$ projects to 0 in J'', $\tau(\tilde{\xi}) - \xi \in J'$. Its image in $(J')_\tau$ is independent of the choice of the lift of ξ and is by definition $\delta(\xi)$. If J and J'' are not assumed to be abelian but J' is central in J, then there is still an exact sequence

$$0 \to (J')^\tau \to J^\tau \to (J'')^\tau \to (J')_\tau,$$

where the connecting homomorphism $\delta\colon (J'')^\tau \to (J')_\tau$ is defined as before.

Let us return to the situation of the compact group H. Let $\mathfrak{z}$ be the center of $\mathfrak{h}$. Then $\mathfrak{t} = \mathfrak{z} \oplus \mathfrak{d}$, where $\mathfrak{d}$ is the Lie algebra of $T \cap DH$. Since by assumption $\sigma|\mathfrak{z}$ has finite order, we can fix a σ-invariant inner product on $\mathfrak{t}$, which we can also take to be invariant under the Weyl group, such that the summands $\mathfrak{z}$ and $\mathfrak{d}$ are orthogonal. The vector space $\mathfrak{d}$ contains the lattice $2\pi i Q_H^\vee$ and is spanned by it. Let Λ be the kernel of the exponential map $\mathfrak{t} \to T$. Thus $\pi_1(H) \cong \Lambda/2\pi i Q_H^\vee$.

The automorphism σ generates a cyclic subgroup of linear automorphisms of $\mathfrak{d}$ which satisfies the hypotheses of Section 6.1 and contains no rotation in case Φ_H has a factor of type A. Orthogonal projection $\pi\colon \mathfrak{t} \to \mathfrak{t}^\sigma$ induces a map $\Lambda_\sigma \to \mathfrak{t}^\sigma$. Since $\Delta_H^\vee$ is a σ-invariant basis of $i\mathfrak{d}$ it follows that $\Delta_H^\vee/\langle\sigma\rangle$ is a basis for $i\mathfrak{d}^\sigma$. Since

σ permutes a basis for $Q^\vee_H$, it follows that $(Q^\vee_H)_\sigma$ is torsion free and that π induces an isomorphism from $(Q^\vee_H)_\sigma$ to a lattice of maximal rank in $i\mathfrak{d}^\sigma$.

There are the corresponding restricted and projection root systems $\Phi^{\rm res}_H(\sigma)$ and $\Phi^{\rm proj}_H(\sigma)$. In particular, $\Delta^\vee_H/\langle\sigma\rangle$ is a set of simple coroots for $\Phi^{\rm proj}_H(\sigma)^\vee$, and the coroot lattice for $\Phi^{\rm proj}_H(\sigma)$ is the lattice $(Q^\vee_H)_\sigma$ spanned by $\{\pi(a^\vee)\}_{\overline{a}\in\Delta_H/\langle\sigma\rangle}$. The root inverse to $\pi(a^\vee)$ is $\epsilon(\overline{a})n_{\overline{a}}\overline{a}|i\mathfrak{t}^\sigma$.

LEMMA 7.3.1. *Let σ be an automorphism of H in normal form. Then $\pi_0(T^\sigma)$ is naturally isomorphic to* $\mathrm{Tor}(\Lambda_\sigma)$. *In particular, if H is simply connected, then T^σ is connected. The fundamental group of $(T^\sigma)^0 = S^\sigma$ is Λ^σ. The coinvariant torus T_σ is connected, and $T_\sigma = \mathfrak{t}^\sigma/\pi(\Lambda)$, where $\pi\colon \mathfrak{t}\to\mathfrak{t}^\sigma$ is orthogonal projection. The natural map $j\colon \Lambda_\sigma \to \mathfrak{t}^\sigma$ induced by π has kernel equal to* $\mathrm{Tor}\,(\Lambda_\sigma)$ *and image contained in $(2\pi i)$ times the dual lattice to the root lattice of $\Phi^{\rm proj}_H(\sigma)$.*

PROOF. Consider the exact sequence

$$0\to\Lambda\to\mathfrak{t}\to T\to 0.$$

Taking the associated long exact cohomology sequence gives

$$0\to\Lambda^\sigma\to\mathfrak{t}^\sigma\to T^\sigma\to\Lambda_\sigma\to\mathfrak{t}_\sigma\to T_\sigma\to 0.$$

By the remarks at the beginning of this section, the kernel of the map $j\colon\Lambda_\sigma\to\mathfrak{t}_\sigma$ is exactly $\mathrm{Tor}(\Lambda_\sigma)$. To see that its image is contained in $(2\pi i)$ times the dual lattice to the root lattice of $\Phi^{\rm proj}_H(\sigma)$, it suffices to show that, if r is a root of $\Phi^{\rm proj}_H(\sigma)$ and $\lambda\in\Lambda_\sigma$ then $\langle r,\lambda\rangle\in 2\pi i\mathbf{Z}$. The root lattice Φ_H has as a basis the roots $\{\epsilon(\overline{a})n_{\overline{a}}\overline{a}\}_{\overline{a}\in\Delta_H/\langle\sigma\rangle}$. We must show that the value of such a basis element on λ lies in $2\pi i\mathbf{Z}$. Let $\{a_1,\dots,a_{n_{\overline{a}}}\}$ be the orbit $\overline{a}$ and lift λ to $\tilde\lambda\in\Lambda$. Then $\langle r,\lambda\rangle = \epsilon(\overline{a})\sum_{i=1}^{n_{\overline{a}}}\langle a_i,\tilde\lambda\rangle\in 2\pi i\mathbf{Z}$.

Since $\mathfrak{t}^\sigma/\Lambda^\sigma$ is connected, it follows that $\pi_0(T^\sigma)\cong\mathrm{Tor}(\Lambda_\sigma)$. If H is simply connected, then $\Lambda = Q^\vee_H$ and $\mathrm{Tor}(\Lambda_\sigma)=0$. The remaining statements are clear. □

We turn now to the corresponding statements in the group H. Let $\widetilde{H}$ be the universal covering group of H. Then there is an exact sequence

$$\{0\}\to\Lambda/2\pi iQ^\vee_H\to\widetilde{H}\to H\to\{1\},$$

where $\Lambda/2\pi iQ^\vee_H$ is a central subgroup of $\widetilde{H}$. The automorphism σ acts on this sequence. Thus, there is an exact sequence

$$\{0\}\to(\Lambda/2\pi iQ^\vee_H)^\sigma\to\widetilde{H}^\sigma\to H^\sigma\to(\Lambda/2\pi iQ^\vee_H)_\sigma\,.$$

Since $\widetilde{H}$ is the product of a vector space and a compact, simply connected group, it follows from 3.1 and 3.5 in [**1**] that $\widetilde{H}^\sigma$ is connected. (In fact, we shall give an independent proof of this fact shortly.) Thus the image of $\widetilde{H}^\sigma$ in H^σ is the identity component of H^σ. Hence the map $\pi_0(H^\sigma)\to(\Lambda/2\pi iQ^\vee_H)_\sigma$ is injective, and its image is finite since $\pi_0(H^\sigma)$ is finite. We have thus proved:

LEMMA 7.3.2. *The function δ induces an injection, which we also denote by δ, from $\pi_0(H^\sigma)$ to* $\mathrm{Tor}\,(\Lambda/2\pi iQ^\vee_H)_\sigma$.

Applying the above to the case $H = T$, there is also a homomorphism from $\pi_0(T^\sigma)$ to $\mathrm{Tor}(\Lambda_\sigma)$. Of course, this is just the isomorphism of Lemma 7.3.1. By the functoriality of the connecting homomorphism δ, we have:

LEMMA 7.3.3. *There is a commutative diagram with exact columns:*

$$\begin{array}{ccc} 0 & & 0 \\ \downarrow & & \downarrow \\ \pi_0(T^\sigma) & \xrightarrow{\cong} & \mathrm{Tor}(\Lambda_\sigma) \\ \downarrow & & \downarrow \\ \pi_0(H^\sigma) & \xrightarrow{\delta} & \mathrm{Tor}\left((\Lambda/2\pi i Q_H^\vee)_\sigma\right). \end{array}$$

Recall that $\mathcal{C}\Phi_H^{\mathrm{proj}}(\sigma)$ is the quotient of the dual of the root lattice of $\Phi_H^{\mathrm{proj}}(\sigma)$ by the coroot lattice $(Q_H^\vee)_\sigma$. It is identified with a finite subgroup of $\mathfrak{d}^\sigma/(Q_H^\vee)_\sigma$. The exact sequence

$$0 \to 2\pi i Q_H^\vee \to \Lambda \to \Lambda/2\pi i Q_H^\vee \to 0$$

yields an exact sequence

$$(2\pi i Q_H^\vee)_\sigma \to \Lambda_\sigma \to (\Lambda/2\pi i Q_H^\vee)_\sigma \to 0.$$

The image in $\mathfrak{t}_\sigma$ of any $\mu \in \Lambda_\sigma$ which maps to a torsion element in $(\Lambda/2\pi i Q_H^\vee)_\sigma$ is contained in $\mathfrak{d}^\sigma$. This gives a homomorphism from $\mathrm{Tor}\left((\Lambda/2\pi i Q_H^\vee)_\sigma\right)$ to the quotient $\mathfrak{d}^\sigma/(2\pi i Q_H^\vee)_\sigma$. The kernel of this homomorphism is the image of $\mathrm{Tor}(\Lambda_\sigma)$ in $\mathrm{Tor}\left((\Lambda/2\pi i Q_H^\vee)_\sigma\right)$ and the image is contained in $\mathcal{C}\Phi_H^{\mathrm{proj}}(\sigma)$. Finally, the map from $\mathrm{Tor}(\Lambda_\sigma)$ in $\mathrm{Tor}\left((\Lambda/2\pi i Q_H^\vee)_\sigma\right)$ is injective by Lemma 7.3.3. We summarize this picture in the following lemma:

LEMMA 7.3.4. *Orthogonal projection from Λ_σ to $\mathfrak{t}_\sigma$ induces an exact sequence*

$$0 \to \mathrm{Tor}(\Lambda_\sigma) \to \mathrm{Tor}\left((\Lambda/2\pi i Q_H^\vee)_\sigma\right) \to \mathcal{C}\Phi_H^{\mathrm{proj}}(\sigma).$$

We have identified $\Lambda_\sigma/\mathrm{Tor}(\Lambda_\sigma)$ with $\pi(\Lambda)$. Moreover $\pi(Q_H^\vee) \cong (Q_H^\vee)_\sigma$. Thus

$$\pi(\Lambda)/\pi(2\pi i Q_H^\vee) \cong (\Lambda/2\pi i Q_H^\vee)_\sigma/\mathrm{Tor}(\Lambda_\sigma).$$

The following lemma, which will be needed in the next chapter, is then clear.

LEMMA 7.3.5. *There is an exact sequence*

$$0 \to \pi(2\pi i Q_H^\vee) \to (\pi(2\pi i Q_H^\vee) \otimes \mathbf{Q}) \cap \pi(\Lambda) \to \mathrm{Tor}\left((\Lambda/2\pi i Q_H^\vee)_\sigma\right)/\mathrm{Tor}(\Lambda_\sigma) \to 0.$$

Fix once and for all a special automorphism σ_0 in normal form with respect to T and A. Let $B \subseteq i\mathfrak{d}^{\sigma_0}$ be the alcove containing the origin determined by the set of simple roots of $\Phi_H^{\mathrm{proj}}(\sigma_0)$ inverse to $\Delta_H^\vee/\langle\sigma_0\rangle$. By Section 3.2, there is an affine action of $\mathcal{C}\Phi_H^{\mathrm{proj}}(\sigma_0)$ on $i\mathfrak{d}^{\sigma_0}$ normalizing $B \subseteq i\mathfrak{d}^{\sigma_0}$. We extend this action to one on all of $i\mathfrak{t}^{\sigma_0}$ by letting $\mathcal{C}\Phi_H^{\mathrm{proj}}(\sigma_0)$ act trivially on $i\mathfrak{z}^{\sigma_0}$. The linearization of the action of $\mathcal{C}\Phi_H^{\mathrm{proj}}(\sigma_0)$ is denoted ν_B.

We now return to the study of the component group of H^σ. By Lemma 7.1.3, we can assume after conjugation that $\sigma = \sigma_s = i_s \circ \sigma_0$, where $s \in T$. By the remarks at the beginning of this section, if s_1 and s_2 are two elements of T which differ by an element of $(\mathrm{Id} - \sigma_0)(T)$, then the subgroups $H^{\sigma_{s_1}}$ and $H^{\sigma_{s_2}}$ are conjugate in H. Furthermore, since the map $S^\sigma \to T/(\mathrm{Id} - \sigma_0)(T)$ is surjective, it is no loss of generality to assume that $s \in S^\sigma$.

LEMMA 7.3.6. *Without changing the conjugacy class of H^{σ_s}, we may assume that there is $\tilde{s} \in i\mathfrak{t}^{\sigma_0}$ with $\exp(2\pi i \tilde{s}) = s$ and such that the projection $\hat{s}$ of $\tilde{s}$ to $i\mathfrak{d}^{\sigma_0}$ lies in the alcove B.*

PROOF. Let $\tilde{s} \in i\mathfrak{t}^{\sigma_0}$ be such that $\exp(2\pi i\tilde{s}) = s$. We can suppose that the projection of $\tilde{s}$ to $i\mathfrak{d}^{\sigma_0}$ lies in an alcove containing the origin. By Part 1 of Lemma 6.4.2, the group $Z_W(\sigma_0)$ acts transitively on the set of all such alcoves, and hence there is a $w \in Z_W(\sigma_0)$ such that the projection of $w(\tilde{s})$ lies in B. Moreover, by Proposition 7.2.5, we may lift w to $g \in N_H(T)^{\sigma_0}$. Then, since $\sigma_0(g) = g$, we have $i_g \circ i_s \circ \sigma_0 \circ i_g^{-1} = i_{w(s)} \circ \sigma_0$, and the fixed subgroups of $i_s \circ \sigma_0$ and $i_g \circ i_s \circ \sigma_0 \circ i_g^{-1} = i_{w(s)} \circ \sigma_0$ are conjugate in H. □

Thus, to analyze the fixed subgroups, we can always assume that the projection $\hat{s}$ of $\tilde{s}$ to $i\mathfrak{d}^{\sigma_0}$ lies in the alcove B. Denote by $\overline{S}^{\sigma_0}$ the quotient of $\mathfrak{t}^{\sigma_0}$ by the image under orthogonal projection of Λ. Note that this terminology is consistent with that of the introduction. The component group of H^{σ_s} only depends on the image of s in $\overline{S}^{\sigma_0}$.

PROPOSITION 7.3.7. *Suppose that σ_0 is special, and let $\sigma_s = i_s \circ \sigma_0$ with $s = \exp(2\pi i\tilde{s})$ for some $\tilde{s} \in i\mathfrak{t}^{\sigma_0}$. Let $w \in Z_W(\sigma_0)$ and let g be a lift of w to $N_{(H^{\sigma_0})^0}(S^\sigma)$. Let $t \in T$ with $t = \exp(2\pi i\tilde{t})$ for some $\tilde{t} \in i\mathfrak{t}$. Then the element tg lies in H^{σ_s} if and only if*

$$(\mathrm{Id} - w)\tilde{s} + (\sigma_0 - \mathrm{Id})\tilde{t} \in \Lambda,$$

if and only if $(\mathrm{Id} - w)\tilde{s} \in \pi(\Lambda)$. Thus, if $\overline{s}$ is the image of s in $T_{\sigma_0} = \overline{S}^{\sigma_0}$, then $w \in N(\sigma_s)$ if and only if $w(\overline{s}) = \overline{s}$ in the induced action of w on T_{σ_0}.

PROOF. Since $\sigma_0(g) = g$, we see that $i_s \circ \sigma_0(tg) = tg$ if and only if

$$s\sigma_0(t)\sigma_0(g)s^{-1} = s\sigma_0(t)gs^{-1} = tg,$$

or equivalently

$$1 = t^{-1}s\sigma_0(t)gs^{-1}g^{-1} = (t^{-1}\sigma_0(t))(s(w \cdot s)^{-1}.$$

Equivalently, in additive notation,

$$(\mathrm{Id} - w)\tilde{s} + (\sigma_0 - \mathrm{Id})\tilde{t} \in \Lambda.$$

Since every lift of w to an element of $N_H(T)$ is of the form tg for some $t \in T$, we see that w lifts to an element of $(N_H(T))^{\sigma_s}$ if and only if $s \equiv ws \bmod (\mathrm{Id} - \sigma_0)(T)$, or equivalently $w(\overline{s}) = \overline{s}$. □

For simplicity, we assume in the remainder of this section that σ_0 has no exceptional orbits, so as to be able to apply Corollary 7.4.2 which will be proved below. A minor modification handles the general case.

PROPOSITION 7.3.8. *Suppose that $\sigma_s = i_s \circ \sigma_0$ with $s = \exp(2\pi i\tilde{s})$ for some $\tilde{s} \in i\mathfrak{t}^{\sigma_0}$ and that the projection $\hat{s}$ of $\tilde{s}$ to $i\mathfrak{d}^{\sigma_0}$ lies in the alcove B. Then the group $\pi_0(H^{\sigma_s})/\pi_0(T^{\sigma_s})$ is isomorphic to the stabilizer of $\hat{s}$ in $\mathcal{C}\Phi_H^{\mathrm{proj}}(\sigma_0)$.*

PROOF. By Proposition 7.3.7, $N(\sigma_s) = \mathrm{Stab}_{Z_W(\sigma_0)}(\overline{s})$ is the stabilizer in the Weyl group of $\Phi_H^{\mathrm{proj}}(\sigma_0)$ of the point $\overline{s}$. By Corollary 7.4.2 below, the Weyl group $W(S^{\sigma_s}, (H^{\sigma_s})^0)$ is the group $Z_W(\sigma_0)(\overline{s})$ generated by reflections in walls defined by the roots in $\Phi_H^{\mathrm{proj}}(\sigma_0)$ which are integral on $\hat{s}$. By Lemma 3.2.2 (and the remarks immediately following it in case $\Phi_H^{\mathrm{proj}}(\sigma_0)$ is not reduced), the quotient $\mathrm{Stab}_{Z_W(\sigma_0)}(\overline{s})/Z_W(\sigma_0)(\overline{s})$ is isomorphic to the stabilizer of $\hat{s}$ under the action of $\mathcal{C}\Phi_H^{\mathrm{proj}}(\sigma_0)$ on B, and this is the statement of the proposition. □

We can now sketch a proof of Theorem 3.4 in [**1**]:

COROLLARY 7.3.9. *If H is simply connected, then, for every automorphism σ of H, the group H^σ is connected.*

PROOF. Since H is simply connected, $\Lambda/2\pi i Q^\vee_H$ is trivial, and hence the group $\pi_0(H^\sigma)/\pi_0(T^\sigma)$ is trivial. By Lemma 7.3.1, $\pi_0(T^\sigma)$ is also trivial. Thus $\pi_0(H^\sigma)$ is trivial, so that H^σ is connected. □

PROPOSITION 7.3.10. *With notation as above, let*

$$p\colon \pi_0(H^{\sigma_s}) \to N(\sigma_s)/W(S^{\sigma_s}, (H^{\sigma_s})^0)$$
$$\delta\colon \pi_0(H^{\sigma_s}) \to \operatorname{Tor}((\Lambda/2\pi i Q^\vee_H)_{\sigma_0})$$

be the homomorphisms defined respectively by the exact sequence of Lemma 7.1.7 and by Lemma 7.3.2, and let ν_B be the linearization of the action of $\mathcal{C}\Phi^{\mathrm{proj}}_H(\sigma_0)$ on B, viewed as a homomorphism from $\operatorname{Tor}((\Lambda/2\pi i Q^\vee_H)_{\sigma_0})$ to $Z_W(\sigma_0)$. Then the image of $\nu_B \circ \delta$ is contained in $N(\sigma_s)$, and $p = \nu_B \circ \delta$ as homomorphisms from $\pi_0(H^{\sigma_s})$ to $N(\sigma_s)/W(S^{\sigma_s}, (H^{\sigma_s})^0)$.

PROOF. Fix a component of H^{σ_s}, and find, by Lemma 7.1.4, an element $g_0 \in N_H(T)^{\sigma_s}$ lying in this component. Let $w \in Z_W(\sigma_0)$ be the image of g_0. Then by definition $w \in N(\sigma_s)$ and $p([g_0]) = w \bmod W(S^{\sigma_s}, (H^{\sigma_s})^0)$. By Proposition 7.2.5, there is a $g \in N_H(T)^{\sigma_0}$ whose image in $Z_W(\sigma_0)$ is also w. Thus $g_0 = tg$ for some $t \in T$. Let $\tilde{g}$ be a lift of g to the covering $\mathfrak{t} \rtimes W$ of $N_H(T)$ and let $\tilde{t}$ be a lift of t to $\mathfrak{t}$. Then the element $\tilde{t}\tilde{g}$ is a lift of tg to $\mathfrak{t} \rtimes W$. The calculation of Proposition 7.3.7 shows that

$$\lambda = (\mathrm{Id} - w)\tilde{s} + (\sigma_0 - \mathrm{Id})\tilde{t} \in \Lambda$$

and identifies $\sigma_0(\tilde{t}\cdot\tilde{g})(\tilde{t}\cdot\tilde{g})^{-1}$ (additively) with λ, up to an element in $(\mathrm{Id} - \sigma_0)(\Lambda)$. It then follows from the definition of δ that $\delta([g_0]) = \delta([tg])$ is the image of λ in $(\Lambda/2\pi i Q^\vee_H)_{\sigma_0}$. Thus, letting π denote the orthogonal projection onto $i\mathfrak{t}^{\sigma_0}$, by Lemma 7.3.4, the image $\overline{\xi}$ of $\delta([tg])$ in $\mathcal{C}\Phi^{\mathrm{proj}}_H(\sigma_0)$ is represented by $\pi(\lambda)$ modulo $\pi(Q^\vee_H)$. Of course, $\pi(\lambda) = \pi(\mathrm{Id} - w)\tilde{s} = (\mathrm{Id} - w)(\hat{s})$. Thus $\xi = \hat{s} - w(\hat{s}) \in i\mathfrak{t}_0^\sigma$ lifts $\overline{\xi}$. In other words, $w(\hat{s}) + \xi = \hat{s}$. Let $B' = w(B) + \xi$; it is an alcove in $i\mathfrak{d}^{\sigma_0}$ for $\Phi^{\mathrm{proj}}_H(\sigma_0)$. Now there is a unique $\gamma \in W_{\mathrm{aff}}(\Phi^{\mathrm{proj}}_H(\sigma_0))$ such that $\gamma(B') = B$ and $\gamma(\hat{s}) = \hat{s}$. Let w' be the Weyl part of γ. It is a product of reflections in walls defined by roots in $\Phi^{\mathrm{proj}}_H(\sigma_0)$ which are integral on $\hat{s}$. By Corollary 7.4.2 below, such walls are walls of roots of H^{σ_s}. Hence $w' \in W(S^{\sigma_s}, (H^{\sigma_s})^0)$. Let φ be the affine transformation $\gamma w + \xi = w'w + \xi'$, where $\xi' \equiv \xi$ mod the coroot lattice of $\Phi^{\mathrm{proj}}_H(\sigma_0)$. Then by construction φ normalizes the alcove B, and so by definition $w'w = \nu_B(\xi') = \nu_B(\xi) = \nu_B(\delta([g_0]))$. Since $w \equiv w'w \bmod W(S^{\sigma_s}, (H^{\sigma_s})^0)$, we see that $\nu_B(\delta([g_0])) \in N(\sigma_s)$ and that $p([g_0]) = w'w = \nu_B(\delta([g_0]))$. This completes the proof. □

COROLLARY 7.3.11. *For $s \in S^\sigma$, let $\sigma_s = i_s \circ \sigma_0$, let $\hat{s}$ be the image of $\tilde{s}$ in $i\mathfrak{d}^{\sigma_0}$, and assume that $\hat{s}$ lies in the alcove B. Then the group $\pi_0(H^{\sigma_s})$ is isomorphic to the stabilizer of $\hat{s}$ in the group $\operatorname{Tor}((\Lambda/2\pi i Q^\vee_H)_{\sigma_0})$. Moreover, for every element $\mu \in \operatorname{Tor}((\Lambda/2\pi i Q^\vee_H)_{\sigma_0})$, there exists an s such that $\mu = \delta([z])$ for some $z \in H^{\sigma_s}$.*

PROOF. Let $\overline{\delta}$ be the composed homomorphism

$$\pi_0(H^{\sigma_s}) \to \operatorname{Tor}((\Lambda/2\pi i Q^\vee_H)_{\sigma_0}) \to \mathcal{C}\Phi^{\mathrm{proj}}_H(\sigma_0).$$

By Proposition 7.3.10, the image of $\overline{\delta}$ is exactly the stabilizer of $\hat{s}$ in $\mathcal{C}\Phi_H^{\mathrm{proj}}(\sigma_0)$. Since the image of δ contains the kernel of the map from $\mathrm{Tor}\left((\Lambda/2\pi i Q_H^\vee)_{\sigma_0}\right)$ to $\mathcal{C}\Phi_H^{\mathrm{proj}}(\sigma_0)$, it follows that the image of δ is the stabilizer of $\hat{s}$ in $\mathrm{Tor}\left((\Lambda/2\pi i Q_H^\vee)_{\sigma_0}\right)$.

The second statement follows immediately from the fact that every affine automorphism of a simplex has a fixed point. □

A slightly less precise result appears as Theorem 9.1 in [**18**].

7.4. The roots of H^σ

Aside from giving a proof of Corollary 7.4.2, used above, this section is a digression whose purpose is to show that the methods developed in this paper can be used to give a description of the Lie algebra of H^σ, which will not be needed in the rest of the paper. Most of these results, stated in a somewhat different language, can be found in III §3 no. 6 of de Siebenthal [**17**] (the main concern of that paper was to show that the conjugacy classes in the disconnected group which is the extension of H by the finite cyclic group generated by the outer automorphism corresponding to σ are given by the alcove B).

As in the previous section, it suffices to consider the case where $\sigma = \sigma_s = i_s \circ \sigma_0$, where σ_0 is a special automorphism in normal form and $s \in S^\sigma$. We let $\tilde{s}$, $\hat{s}$, and B have the same meaning as in Lemma 7.3.6.

LEMMA 7.4.1. *Let σ_0 be a special automorphism and suppose that the action of σ_0 on Δ_H has no exceptional orbits. Let $\tilde{s} \in \mathfrak{t}^{\sigma_0}$, and set $s = \exp(2\pi i \tilde{s})$. Then the roots of H^{σ_s} with respect to the maximal torus S^{σ_0} are*

$$\{\overline{a} \in \Phi_H^{\mathrm{res}}(\sigma_0) \big| \; n_{\overline{a}}\langle \overline{a}, \tilde{s}\rangle \in \mathbf{Z}\rangle\}.$$

PROOF. Since σ_0 has no exceptional orbits, it follows from Lemma 6.4.1 that $\Phi_H^{\mathrm{res}}(\sigma_0)$ is a reduced root system. Thus, by Lemma 7.2.5 each $\overline{a}$ in $\Phi_H^{\mathrm{res}}(\sigma_0)$ is a root of of H^{σ_0} with respect to $\mathfrak{t}^{\sigma_0}$.

The roots of H^{σ_s} with respect to $\mathfrak{t}^{\sigma_0}$ are a subset of $\Phi_H^{\mathrm{res}}(\sigma_0)$. Let $\overline{a} \in \Phi_H^{\mathrm{res}}(\sigma_0)$ be an orbit of order $n_{\overline{a}}$. Let $\{a_1, \ldots, a_{n_{\overline{a}}}\}$ be this orbit, ordered so that $\sigma_0 \cdot a_i = a_{i+1}$ (by convention $a_{n_{\overline{a}}+1} = \overline{a}_1$). For each i let $\mathfrak{h}^{a_i} \subseteq \mathfrak{h} \otimes \mathbf{C}$ be the root space for a_i. Since $\overline{a}$ is a root of H^{σ_0}, it follows that the action of σ_0 identifies the root space $\mathfrak{h}^{a_i}$ with $\mathfrak{h}^{a_{i+1}}$ in such a way that the action of $\sigma_0^{n_{\overline{a}}}$ is the identity on each of these root spaces. Since $s \in S^{\sigma_0}$, we have $s^{a_i} = s^{a_{i+1}}$ for all $i \leq n_{\overline{a}}$. Thus, $s^{n_{\overline{a}}a_1} = 1$ if and only if the action of $(s\sigma_0)^{n_{\overline{a}}}$ is trivial on all these root spaces if and only if $\overline{a}$ is a root of H^{σ_s}. Of course $s^{n_{\overline{a}}a_1} = 1$ if and only if $n_{\overline{a}}\langle \overline{a}, \tilde{s}\rangle \in \mathbf{Z}$. □

COROLLARY 7.4.2. *With σ_0, $\tilde{s}$ and σ_s as in the previous lemma, the roots of H^{σ_s} are exactly those of $\Phi_H^{\mathrm{res}}(\sigma_0)$ that correspond under the bijection given in Corollary 7.2.6 to the roots of $\Phi_H^{\mathrm{proj}}(\sigma_0)$ which take integral values on $\tilde{s}$.*

PROOF. Since σ_0 has no exceptional orbits, $\Phi_H^{\mathrm{proj}}(\sigma_0)$ is a reduced root system and the root of this system corresponding to $\overline{a} \in \Phi_H^{\mathrm{res}}(\sigma_0)$ is $n_{\overline{a}}\overline{a}$. Given this the result is immediate from the previous lemma. □

Suppose that σ_0 is nontrivial, that Φ is irreducible, and that there are no exceptional orbits. It follows that Φ is simply laced. By Lemma 6.2.7 the systems $\Phi_H^{\mathrm{res}}(\sigma_0)$ and $\Phi_H^{\mathrm{proj}}(\sigma_0)$ are inverse to each other. The roots of H^{σ_s} are then the elements of $\Phi_H^{\mathrm{res}}(\sigma_0)$ such that the corresponding inverse coroot is integral on $\tilde{s}$, or equivalently on $\hat{s}$. Since $\hat{s}$ lies in the alcove B, a set of simple roots for H^{σ_s} is obtained as follows: take the extended coroot diagram for $\Phi_H^{\mathrm{res}}(\sigma_0)$, in other words the extended root diagram for $\Phi_H^{\mathrm{proj}}(\sigma_0)$. Let $\Phi_H^{\mathrm{proj}}(\sigma_0)(\hat{s})$ be the set of elements of $\Phi_H^{\mathrm{proj}}(\sigma_0)$ such that the corresponding wall of B contains $\hat{s}$. Then the set of elements of $\Phi_H^{\mathrm{res}}(\sigma_0)$ which are inverse to an element of $\Phi_H^{\mathrm{proj}}(\sigma_0)(\hat{s})$ is a set of simple roots for H^{σ_s}. The possible diagrams obtained in this way will describe all of the possible root systems for H^{σ_s}. A similar result holds if there are exceptional orbits, with a slightly more involved proof. In this case, the root and coroot systems are both of type BC_n, and thus are abstractly isomorphic, and the procedure for finding the possible root systems of the H^{σ_s} is again to take the extended coroot diagram, choose a proper subdiagram, and then pass to the inverse system. Taken together, these results generalize a theorem of Kac [**9**] which dealt with the case of finite order automorphisms σ.

7.5. The case of c-pairs

We return to the notation of the previous chapters, so that G is a simple and simply connected group with maximal torus T, and A is an alcove in $i\mathfrak{t}$. Let $c \in \mathcal{C}G$, and suppose that (x, y) is a c-pair. After conjugation we can assume that (x, y) is in normal form, so that $x = \exp(2\pi i\tilde{x})$ for some $\tilde{x} \in A^c$. We will apply the results above to the compact group $H = Z(x)$ and the automorphism of $Z(x)$ defined by conjugation by y. Thus T is a maximal torus of $Z(x)$ and i_y is an automorphism in normal form of $Z(x)$ whose restriction to T is w_c. For future reference, let us record the following definition:

DEFINITION 7.5.1. A c-pair (x, y_0) in normal form such that i_{y_0} is a special automorphism of $Z(x)$ is called a *special c-pair.*

If (x, y) is a c-pair in normal form, T is a maximal torus for $Z(x)$ and that a set of simple roots $\widetilde{I}(x)$ for $Z(x)$ is given by the subset of $\widetilde{\Delta}$ taking integral values on $\tilde{x}$. Let $Q^\vee(x)$ be the lattice generated by $\widetilde{I}(x)$. It is the coroot lattice for the derived subgroup of $Z(x)$, and thus is the lattice denoted $Q_H^\vee$ above. Conjugation by y induces an automorphism of $Z(x)$ normalizing $\mathfrak{t}$ and $A \subseteq \mathfrak{t}$. Its action on $\mathfrak{t}$ is given by the Weyl element w_c. This Weyl element normalizes $\widetilde{\Delta}$ and $\widetilde{I}(x)$. Let $\widetilde{\Delta}_c$ and $\widetilde{I}_c(x)$ be the respective quotients of these sets by the action of w_c. The elements of $\widetilde{\Delta}_c \subseteq \mathfrak{t}^{w_c}$ satisfy the single relation $\sum_{\overline{a}\in\widetilde{\Delta}_c} g_{\overline{a}}\pi(a^\vee) = 0$.

We turn now to a description of $\mathrm{Tor}\,((Q^\vee/Q^\vee(x))_{w_c})$:

LEMMA 7.5.2. *$\mathrm{Tor}\,((Q^\vee/Q^\vee(x))_{w_c})$ is a cyclic group of order $n = \gcd\{\, g_{\overline{a}} : \overline{a} \in \widetilde{\Delta}_c - \widetilde{I}_c(x)\,\}$, generated by*

$$\zeta = \frac{1}{n}\sum_{\overline{a}\in\widetilde{\Delta}_c-\widetilde{I}_c(x)} g_{\overline{a}}\pi(a^\vee) = -\sum_{\overline{a}\in\widetilde{I}_c(x)} \frac{g_{\overline{a}}}{n}\pi(a^\vee) \in Q^\vee.$$

PROOF. We have an exact sequence

$$0 \to \mathbf{Z}\left(\sum_{a\in\widetilde{\Delta}} g_a a^\vee\right) \to \bigoplus_{a\in\widetilde{\Delta}} \mathbf{Z}(a^\vee) \to Q^\vee \to 0.$$

Since $Q^\vee(x) = \bigoplus_{a\in\widetilde{I}(x)} \mathbf{Z}(a^\vee)$, we see that there is an exact sequence

$$0 \to \mathbf{Z}\left(\sum_{a\in\widetilde{\Delta}-\widetilde{I}(x)} g_a a^\vee\right) \to \bigoplus_{a\in\widetilde{\Delta}-\widetilde{I}(x)} \mathbf{Z}(a^\vee) \to Q^\vee/Q^\vee(x) \to 0.$$

Conjugation by y induces an action on this sequence. On the second term it is the action induced by the permutation action of $\widetilde{\Delta}-\widetilde{I}(x)$ with quotient $\widetilde{\Delta}_c - \widetilde{I}_c(x)$. Hence, taking coinvariants yields an exact sequence

$$\mathbf{Z}\left(\sum_{\overline{a}\in\widetilde{\Delta}_c-\widetilde{I}_c(x)} g_{\overline{a}}\pi(a^\vee)\right) \to \bigoplus_{\overline{a}\in\Delta_c-\widetilde{I}_c(x)} \mathbf{Z}(\pi(a^\vee)) \to \left(Q^\vee/Q^\vee(x)\right)_{w_c} \to 0.$$

Clearly, then, the torsion subgroup of $\left(Q^\vee/Q^\vee(x)\right)_{w_c}$ is as claimed. Since $n|g_{\overline{a}}$ for all $\overline{a}\in\widetilde{\Delta}_c-\widetilde{I}_c(x)$, we see that $\zeta\in Q^\vee$. The two expressions for ζ are equal since, in $Q^\vee$,we have the relation $\sum_{\overline{a}\in\widetilde{\Delta}} g_{\overline{a}}\pi(a^\vee)=0$. □

COROLLARY 7.5.3. *The group $\pi_0(Z(x,y))$ is cyclic of order dividing*

$$n = \gcd\{\, g_{\overline{a}} : \overline{a}\in\widetilde{\Delta}_c - \widetilde{I}_c(x)\,\}.$$

The order of $\pi_0(T^{w_c})$ is $n_0 = \gcd\{g_{\overline{a}} : \overline{a}\in\widetilde{\Delta}_c\}$. For every c-pair in normal form, the map from $\pi_0(T^{w_c})$ to $\pi_0(Z(x,y))$ is an injection whose image is the cyclic subgroup of $\pi_0(Z(x,y))$ of order n_0.

PROOF. By Lemma 7.3.2, $\pi_0(Z(x,y))$ is a subgroup of $\mathrm{Tor}\left((Q^\vee/Q^\vee(x))_{w_c}\right)$, and hence it is cyclic of order dividing $n=\gcd\{\,g_{\overline{a}} : \overline{a}\in\widetilde{\Delta}_c-\widetilde{I}_c(x)\,\}$. By Lemma 7.3.1, $\pi_0(T^{w_c})$ is isomorphic to $\mathrm{Tor}(Q^\vee)_{w_c}$, which has order $n_0=\gcd\{g_{\overline{a}} : \overline{a}\in\widetilde{\Delta}_c\}$ by applying Lemma 7.5.2 in the case where $\widetilde{I}_c(x)=\emptyset$. The last statement then follows from Corollary 7.3.3. □

We now deal with the case where the action of w_c on $\widetilde{\Delta}$ has an exceptional orbit as defined in Section 6.7.

LEMMA 7.5.4. *Let G be simple and let $c\in\mathcal{C}G$. If the action of w_c on $\widetilde{\Delta}$ has an exceptional orbit, then the integers $\{g_{\overline{a}}\}$ are all equal and the number of orbits is at least 2.*

PROOF. Suppose that w_c acts on $\widetilde{\Delta}$ interchanging two roots $a_1^\vee, a_2^\vee$ which are not orthogonal. The extended Dynkin diagram $\widetilde{D}^\vee(G)$ is thus symmetric about the bond connecting the nodes corresponding to a_1 and a_2. This symmetry implies that there are three possible types of diagrams to consider: (i) $\widetilde{D}$ is simply laced and has no trivalent vertices; (ii) $\widetilde{D}$ is simply laced and has two trivalent vertices; and (iii) $\widetilde{D}$ has two double bonds. In the first case G is isomorphic to $SU(2n+1)$ for some $n\geq 1$. This case is ruled out since there are no exceptional orbits for the action of the center of $SU(2n+1)$ on the extended Dynkin diagram of A_{2n}. In the second case, the subdiagram of $\widetilde{D}$ which contains the chain connecting the trivalent vertices together with all nodes adjacent to the trivalent vertices is the extended Dynkin diagram for D_{2n+1}, and hence there is a non-trivial linear relation between the elements of $\widetilde{\Delta}$ corresponding to the nodes of this subdiagram. Since any proper subset of $\widetilde{\Delta}$ is linearly independent, it follows that this is the entire extended diagram for G. Thus, G is of type D_{2n+1} for some $n\geq 2$. Direct examination of the action of the center in this case shows that c is a generator of the center and that the integers $\{g_{\overline{a}}\}$ are all equal to 4. In the third case, $\widetilde{D}$ has no trivalent vertices and hence is a chain. Since the highest root is a long root and since according to Lemma 3.3.1 it must be at one end of the chain, it follows that the a_1 and a_2 are short roots. The subchain that contains a_1 and a_2 and contains one long root on each side of a_1 is the extended diagram for C_{2n+1} for some $n \geq 1$. As before, since there is a nontrivial linear relation between these roots, it follows that G is of type C_{2n+1}. Direct inspection shows that the $\{g_{\overline{a}}\}$ are all equal to 2 in this case. □

PROPOSITION 7.5.5. *Suppose that all the $g_{\overline{a}}$ are equal, say to n. Then for every c-pair (x,y) in normal form the group $Z(x,y)$ has n components. The inclusion $T^{w_c} \to Z(x,y)$ induces a bijection on the group of components, which is isomorphic to* $\mathrm{Tor}(Q^\vee/Q^\vee(x))_{w_c}$.

PROOF. By Lemma 7.5.3, $\pi_0(Z(x,y))$ is a subgroup of $\mathrm{Tor}(Q^\vee/Q^\vee(x))_{w_c}$, which is cyclic of order $n = n_0$, and the map from $\pi_0(T^{w_c})$ to $\pi_0(Z(x,y))$ is injective. Since the order of $\pi_0(T^{w_c})$ is also n, it follows that the natural maps $\pi_0(T^{w_c}) \to \pi_0(Z(x,y)) \to \mathrm{Tor}(Q^\vee/Q^\vee(x))_{w_c}$ are all isomorphisms. □

7.6. Variation of $\pi_0(Z(x,y))$ as x varies

PROPOSITION 7.6.1. *Let $\tilde{x}, \tilde{x}'$ be points of A^c. Let $x = \exp(2\pi i\tilde{x})$, resp. $x' = \exp(2\pi i\tilde{x}')$. Suppose $Z(x') \subseteq Z(x)$. Then the homomorphism on fundamental groups $Q^\vee/Q^\vee(x') \to Q^\vee/Q^\vee(x)$ induced by the inclusion $Z(x') \subseteq Z(x)$ descends to a map on coinvariants which gives an injection*

$$\mathrm{Tor}\left((Q^\vee/Q^\vee(x'))_{w_c}\right) \subseteq \mathrm{Tor}\left((Q^\vee/Q^\vee(x))_{w_c}\right).$$

PROOF. The groups $Z(x)$ and $Z(x')$ are connected and both have T as a maximal torus. Since $Z(x') \subseteq Z(x)$, $\widetilde{I}(x') \subseteq \widetilde{I}(x)$, and hence $Q^\vee(x') \subseteq Q^\vee(x)$. The natural surjection $p\colon Q^\vee/Q^\vee(x') \to Q^\vee/Q^\vee(x)$ is the map on fundamental groups induced by the inclusion $Z(x') \subseteq Z(x)$. It follows immediately from the description of the torsion subgroup and its generator given in Lemma 7.5.2 that, if $\mathrm{Tor}((Q^\vee/Q^\vee(x))_{w_c})$ is cyclic of order n and $\mathrm{Tor}((Q^\vee/Q^\vee(x'))_{w_c})$ is cyclic of order n', then $n'|n$ and that the map $p_{w_c}\colon (Q^\vee/Q^\vee(x'))_{w_c} \to (Q^\vee/Q^\vee(x))_{w_c}$ induced by p on coinvariants sends a generator of $\mathrm{Tor}((Q^\vee/Q^\vee(x'))_{w_c})$ to n/n' times a generator of $\mathrm{Tor}((Q^\vee/Q^\vee(x))_{w_c})$. Thus it is injective on the torsion subgroup. □

COROLLARY 7.6.2. *Let (x,y) and (x',y') be c-pairs in normal form. Suppose $Z(x') \subseteq Z(x)$ and that $Z(x',y') \subseteq Z(x,y)$. Then the inclusion $Z(x',y') \to Z(x,y)$ induces an injective homomorphism $\pi_0(Z(x',y')) \to \pi_0(Z(x,y))$.*

PROOF. This is immediate from Proposition 7.6.1 and Corollary 7.5.3. □

CHAPTER 8

C-triples

Let $C = (c_{ij})_{1 \le i,j \le 3}$ be an antisymmetric matrix of elements of $\mathcal{C}G$. Let $\langle C \rangle \subseteq \mathcal{C}G$ be the subgroup generated by all the entries c_{ij} of C. We wish to study the moduli space of conjugacy classes of ordered C-triples in G. Quite generally, suppose that $\mathbf{x} = (x_1, x_2, x_3)$ is an ordered triple in G^3. For an integral 3×3 matrix M, we set $\mathbf{x}^M = (x_1', x_2', x_3')$, where $x_i' = x_1^{m_{1i}} x_2^{m_{2i}} x_3^{m_{3i}}$. We have the following straightforward lemma:

LEMMA 8.0.3. *Suppose that M is unimodular. The map $\mathbf{x} \mapsto \mathbf{x}^M$ defines an homeomorphism from the moduli space of C-triples to the moduli space of C'-triples, where $C' = (c_{ij}')_{1 \le i,j \le 3}$ and $c_{ij}' = \prod_{r,s} c_{rs}^{m_{ri} \cdot m_{sj}}$. Moreover, for an appropriate choice of a unimodular M, either $\langle C \rangle = \langle C' \rangle$ is cyclic and we can assume that $c_{13}' = c_{23}' = 1$ or $G = Spin(4n)$ and $\langle C \rangle = \langle C' \rangle$ is not cyclic, and we can assume that $c_{12}', c_{23}', c_{13}'$ are the three nontrivial elements of $\langle C \rangle$.*

In case $\langle C \rangle$ is cyclic, we assume that $c_{13}' = c_{23}' = 1$ and set $c_{12}' = c$. In other words, (x, y) is a c-pair and $z \in Z(x, y)$. We shall call such a triple a *c-triple*. For the rest of this chapter, we shall study c-triples, postponing the remaining case until later. We let $\mathcal{T}_G(c)$ denote the moduli space of conjugacy classes of c-triples in G.

8.1. c-triples of rank zero

8.1.1. The order of a c-triple.

DEFINITION 8.1.1. We define the *order* of a c-triple $\mathbf{x} = (x, y, z)$ to be the order of $[z]$ in $\pi_0(Z(x, y))$. Clearly, the order is a conjugacy class invariant.

It is easy to check, using the results of this chapter, that this definition coincides with the previous definition given in Chapter 5 in case $c = 1$. It follows from Corollary 8.2.5 below that the order is constant on connected components of $\mathcal{T}_G(c)$. Thus we define the *order* of a component X of $\mathcal{T}_G(c)$ to be the order of $\mathbf{x}$, where $\mathbf{x}$ is any c-triple whose conjugacy class lies in X.

8.1.2. The w_c-action on $\widetilde{\Delta}$.

LEMMA 8.1.2. *Let (x, y, z) be a c-triple of order k and rank zero such that the c-pair (x, y) is in normal form. Then:*

1. *$Z^0(x, y)$ is the torus $S^{w_c} = (T^{w_c})^0$ and the action of i_z on S^{w_c} has isolated fixed points.*
2. *If $k = 1$, then (x, y) is a rank zero c-pair in G, which is of type A, and c generates the center of G.*
3. *A c-triple (x, y, z') is conjugate to (x, y, z) if and only if z and z' lie in the same component of $Z(x, y)$.*

PROOF. The fact that (x,y,z) is of rank zero means that $Z(x,y)^z$ is finite. If $Z^0(x,y)$ is not a torus, then, by **[17]** II §2, $\left(Z^0(x,y)\right)^z \subseteq Z(x,y,z)$ is of positive dimension. Thus $Z^0(x,y)$ is a torus, and since S^{w_c} is a maximal torus of $Z^0(x,y)$ they are equal.

If $k=1$, then $z \in Z^0(x,y)$ which by Part 1 is a torus. Thus, the action of i_z on $Z^0(x,y)$ is trivial. By Part 1 this implies that $Z^0(x,y)$ is a point, which means that (x,y) is a c-pair of rank zero. The statements about G and c now follow from Proposition 4.1.1.

By Part 1 and Corollary 7.5.3, $Z(x,y)$ is an extension of a cyclic group by a torus. Thus (x,y,z') is conjugate to (x,y,z) only if z' is in the same component as z. The converse follows easily from the fact that i_z acts on $S^{w_c} = Z^0(x,y)$ with isolated fixed points, and hence that $S^{w_c} = (\mathrm{Id} - i_z)S^{w_c}$. □

Now let us eliminate one exceptional case.

LEMMA 8.1.3. *Let G be simple and let $c \in \mathcal{C}G$. If there is a c-triple of rank zero in G, then the action of w_c on $\widetilde{\Delta}$ has no exceptional orbits.*

PROOF. Suppose that (x,y,z) is a c-triple of rank zero in G. Then we know that $Z^0(x,y) = S^{w_c}$. If there is an exceptional orbit for the action of w_c on $\widetilde{\Delta}$, then it follows from Lemma 7.5.4 and Proposition 7.5.5 that $Z(x,y) = T^{w_c}$. Consequently, the action of any $z \in Z(x,y)$ on S^{w_c} is trivial. Thus the rank of $Z(x,y,z)$ is equal to that of $Z(x,y)$. But the rank of $Z(x,y)$ is one less than the number of orbits and hence is positive by Lemma 7.5.4. This contradicts the fact that (x,y,z) is of rank zero. □

8.1.3. The divisibility of the $g_{\overline{a}}$. Suppose that (x,y,z) is a c-triple of order k and rank zero and that the c-pair (x,y) is in normal form. Recall that, by Lemma 7.5.2, $\mathrm{Tor}\left((Q^\vee/Q^\vee(x))_{w_c}\right)$ is a cyclic group of order $n = \gcd\{\, g_{\overline{a}} : \overline{a} \in \widetilde{\Delta}_c - \widetilde{I}_c(x)\,\}$, and is generated by

$$\zeta = \frac{1}{n}\sum_{\overline{a}\in\widetilde{\Delta}_c - \widetilde{I}_c(x)} g_{\overline{a}}\pi(a^\vee) = -\sum_{\overline{a}\in\widetilde{I}_c(x)} \frac{g_{\overline{a}}}{n}\pi(a^\vee) \in \pi(Q^\vee).$$

Let $\Phi(x) \subseteq \Phi$ be the set of roots annihilating x and let $i\mathfrak{d}$ be the subspace of $i\mathfrak{t}$ spanned by the coroots inverse to the elements of $\Phi(x)$. Then w_c acts on $i\mathfrak{d}$ and on $\Phi(x)$. Let $\Phi_c(x)$ be the root system $\Phi^{\mathrm{proj}}_{Z(x)}(w_c)$ as defined in the previous chapter. Recall that the coroots inverse to the simple roots of $\Phi_c(x)$ are given by $\{\,\pi(a^\vee) : \overline{a} \in \widetilde{I}_c(x)\,\}$. Fix an alcove B containing the origin in $i\mathfrak{d}^{w_c}$ for the root system $\Phi_c(x)$.

We can write $y = sy_0$ where y_0 is special and $s \in S^{w_c}$. Let $s = \exp(2\pi i\tilde{s})$, where $\tilde{s} \in i\mathfrak{t}^{w_c}$. After a further conjugation, by Lemma 7.3.6, we can assume that $\tilde{s} \in i\mathfrak{t}^{w_c}$ projects to $\hat{s} \in i\mathfrak{d}^{w_c}$ lying in the alcove B. The element z defines a class $[z] \in \pi_0(Z(x,y))$. Let μ_z be the image of $\delta([z])$ in $\mathrm{Tor}\left((Q^\vee/Q^\vee(x))_{w_c}\right)/\mathrm{Tor}(Q^\vee_{w_c}) \subseteq \mathcal{C}\Phi_c(x)$, where δ is the homomorphism of Lemma 7.3.2. By Lemma 7.3.5, we may lift μ_z to an element $\tilde{\mu}_z$ of $(Q^\vee(x)_{w_c}) \otimes \mathbf{Q}$, well-defined modulo $Q^\vee(x)_{w_c}$. Write

$$\tilde{\mu}_z = \sum_{\overline{a}\in\widetilde{I}_c(x)} \lambda_{\overline{a}}\pi(a^\vee). \tag{8.1}$$

PROPOSITION 8.1.4. *Let (x,y,z) be a c-triple of order k and rank zero, and suppose that the c-pair (x,y) is in normal form. With assumptions and notation as in the previous paragraph,*

(1) *The group $Z^0(x,y_0)$ is semisimple and hence $\mathfrak{d}^{w_c} = \mathfrak{t}^{w_c}$.*
(2) *The element $\tilde{s} = \hat{s}$ is a barycenter of the alcove B.*
(3) *In Equation 8.1 no coefficient $\lambda_{\overline{a}}$ is integral.*
(4) *For every set of simple roots of $\Phi_c(x)$, the coefficients of $\tilde{\mu}_z$, written as a linear combination of the inverse coroots, are all non-integral.*
(5) *The element $x = \exp(2\pi i v)$ where v is the vertex of A^c opposite to a face $\{\overline{a} = 0\}$ for some $\overline{a} \in \widetilde{I}_c(x)$.*
(6) *The order k divides $g_{\overline{a}}$.*

PROOF. Since (x,y,z) is of rank zero, Lemma 8.1.2 implies that $Z^0(x,y) = S^{w_c}$ and that conjugation by z normalizes S^{w_c} and its action on this torus has isolated fixed points. Since $Z^0(x,y)$ is abelian, its Weyl group is trivial. Thus, by Lemma 7.1.6 there is a well-defined action of the element $[z] \in \pi_0(Z(x,y))$ on $\mathfrak{t}^{\sigma}$, and, by Proposition 7.3.10, it is given as the Weyl element $w = \nu_B(\mu_z)$. The element w preserves $\mathfrak{t}^{w_c}$ and fixes only the origin there. By Proposition 7.2.5, w is an element of the Weyl group of $Z^0(x,y_0)$. It follows that $Z^0(x,y_0)$ has finite center and thus is semisimple. This proves (1).

Since $w = \nu_B(\mu_z)$, the affine action of $\mu_z \in \mathcal{C}\Phi_c(x)$ on the alcove B has isolated fixed points, and hence fixes only the barycenter of B. Since μ_z fixes $\tilde{s} = \hat{s}$, by Proposition 7.3.8, it follows that $\tilde{s}$ is the barycenter of B, proving Part (2). Moreover, by Proposition 3.5.1, no coefficient $\lambda_{\overline{a}}$ of $\tilde{\mu}_z \in (Q^\vee(x)_{w_c}) \otimes \mathbf{Q}$ is integral, proving (3).

Any two sets of simple roots of $\Phi_c(x)$ are related by an element of the Weyl group of $\Phi_c(x)$. Since the Weyl group acts trivially on the center, and since $\tilde{\mu}_z$ projects to an element of the center, Part (4) follows.

Since $Z^0(x,y_0)$ is semisimple, $\widetilde{I}_c(x)$ has cardinality equal to the dimension of $\mathfrak{t}^{w_c}$. Thus x is the image under the exponential map of a vertex $\tilde{x}$ of A^c, opposite the face $\{\overline{a} = 0\}$, say. By Lemma 7.5.2, $\mathrm{Tor}((Q^\vee/Q^\vee(x))_{w_c})$ is cyclic of order $g_{\overline{a}}$. Since the class of z is an element of order k in $\pi_0(Z(x,y))$, and since δ is injective by Lemma 7.3.2, $\delta([z])$ is an element of order k in $\mathrm{Tor}((Q^\vee/Q^\vee(x))_{w_c})$. Consequently, $k | g_{\overline{a}}$. □

To see the relationship between k and the remaining $g_{\overline{b}}$, we use the following lemma:

LEMMA 8.1.5. *Let $x = \exp(2\pi i\tilde{x})$, where $\tilde{x} \in A^c$. Let $\mu \in (Q^\vee/Q^\vee(x))_{w_c}$ have order k. Let $\tilde{\mu} \in (Q^\vee(x)_{w_c}) \otimes \mathbf{Q}$ be an element lifting the image $\overline{\mu}$ of μ in* $\mathrm{Tor}((Q^\vee/Q^\vee(x))_{w_c})/\mathrm{Tor}(Q^\vee_{w_c}) \subseteq \mathcal{C}\Phi_c(x)$. *Then no coefficient of $\tilde{\mu}$ is integral if and only if, for all $\overline{b} \in \widetilde{I}_c(x)$, $k \nmid g_{\overline{b}}$.*

PROOF. Let $n = \gcd\{\, g_{\overline{a}} : \overline{a} \in \widetilde{\Delta}_c - \widetilde{I}_c(x) \,\}$. By Lemma 7.5.2,

$$\zeta = -\sum_{\overline{b} \in \widetilde{I}_c(x)} (g_{\overline{b}}/n)\pi(b^\vee)$$

is a generator for $(Q^\vee/Q^\vee(x))_{w_c}$. Thus $k|n$ and we can write $\mu = \ell\zeta$ for some integer ℓ of the form tn/k, where t and k are relatively prime. A representative for

$\tilde{\mu}$ is then given by

$$\ell\zeta = -\sum_{\overline{b}\in\widetilde{I}_c(x)} \frac{tg_{\overline{b}}}{k}\pi(b^\vee).$$

Clearly, then, no coefficient of $\tilde{\mu}$ is integral if and only if, for all $\overline{b}\in\widetilde{I}_c(x)$, $k\nmid g_{\overline{b}}$. □

COROLLARY 8.1.6. *Let (x,y,z) be a rank zero c-triple of order k. Then there is a unique $\overline{a}\in\widetilde{\Delta}_c$ such that $k|g_{\overline{a}}$.*

PROOF. This is immediate from the previous lemma and Parts (3) and (6) of the previous proposition. □

8.1.4. Classification of rank zero c-triples. We can now give the classification of rank zero c-triples.

PROPOSITION 8.1.7. *Let G be simple.*

(1) *There is a c-triple of rank zero and order k in G if and only if k divides exactly one of the $g_{\overline{a}}$.*
(2) *Suppose that $g_{\overline{a}}$ does not divide $g_{\overline{b}}$ for $\overline{b}\neq\overline{a}$. Let (x,y,z) be a c-triple of rank zero and order $g_{\overline{a}}$ and let (x',y',z') be a rank zero c-triple whose order k divides $g_{\overline{a}}$. Then (x',y',z') is conjugate to (x,y,z^ℓ) for a unique ℓ such that $1\leq\ell<g_{\overline{a}}$ and moreover $k=g_{\overline{a}}/\gcd(\ell,g_{\overline{a}})$.*

PROOF. The only if part of the first statement is Corollary 8.1.6.

Conversely, suppose that exactly one of the $g_{\overline{a}}$ is divisible by k. It follows from Lemma 7.5.4 that w_c has no exceptional orbits. Choose $\tilde{x}\in A^c$ so that, for $x=\exp(2\pi i\tilde{x})$, we have $\widetilde{I}_c(x)=\widetilde{\Delta}_c-\{\overline{a}\}$. By Lemma 7.5.2, the order of $\mathrm{Tor}\left((Q^\vee/Q^\vee(x))_{w_c}\right)$ is divisible by $g_{\overline{a}}$ and hence is divisible by k. Let y_0 be such that (x,y_0) is a special c-pair in normal form. Let $\tilde{s}\in i\mathfrak{t}^{w_c}$ be the barycenter of the alcove $B\subseteq\mathfrak{t}^{w_c}$ for the root system $\Phi_c(x)$ and let $s=\exp(2\pi i\tilde{s})$. We set $y=sy_0$. Fix an element $\mu\in\mathrm{Tor}\left((Q^\vee/Q^\vee(x))_{w_c}\right)$ of order k. Since $\tilde{s}$ is the barycenter of B, μ fixes $\tilde{s}$. Hence by Corollary 7.3.11, which was proved under the assumption of no exceptional orbits, there is a $z\in Z(x,y)$ whose image in $\mathrm{Tor}\left((Q^\vee/Q^\vee(x))_{w_c}\right)$ under the map $\pi_0(Z(x,y))\to\mathrm{Tor}\left((Q^\vee/Q^\vee(x))_{w_c}\right)$ is μ. Since $\tilde{s}$ is a regular element for $\Phi_c(x)$, it follows that $Z^0(x,y)=S^{w_c}$. Lift the image of μ in $\mathrm{Tor}\left((Q^\vee/Q^\vee(x))_{w_c}\right)/\mathrm{Tor}(Q^\vee)_{w_c}$ to an element $\tilde{\mu}\in(Q^\vee(x))\otimes\mathbf{Q}$, $\tilde{\mu}=\sum s_{\overline{a}}\pi(a^\vee)$. By construction, no coefficient $s_{\overline{a}}$ of a $\pi(a^\vee)\in\widetilde{I}_c(x)$ is integral. Hence, by Proposition 3.5.1, μ acts with isolated fixed points on S^{w_c}. It follows that the conjugation action of z on S^{w_c} also has isolated fixed points, and hence (x,y,z) is of rank zero. Clearly, it is of order k. This proves (1).

To see (2), suppose that $g_{\overline{a}}$ does not divide $g_{\overline{b}}$ for $\overline{b}\neq\overline{a}$. Let (x,y,z) be a c-triple of rank zero and order $g_{\overline{a}}$ and let (x',y',z') be a rank zero c-triple whose order k divides $g_{\overline{a}}$. After conjugation we can assume that (x,y) and (x',y') are c-pairs in normal form. Then $y'\in S^{w_c}\cdot y$. Each of x and x' is $\exp(2\pi iv)$ where v is the vertex of A^c opposite the face $\{\overline{a}=0\}$ for the unique $\overline{a}\in\widetilde{\Delta}_c$ for which $k|g_{\overline{a}}$. Thus, $x=x'$.

By Lemma 7.2.2 there is $y_0\in S^{w_c}\cdot y$ such that (x,y_0) is a special c-pair in normal form. We write $y=sy_0$ and $y'=s'y_0$ for elements $s,s'\in S^{w_c}$ which are the images of $\tilde{s}$ and $\tilde{s}'$ in $i\mathfrak{t}^{w_c}$. It follows from Proposition 8.1.4 that $\tilde{s}$ and $\tilde{s}'$ are barycenters for the alcove decomposition of $i\mathfrak{t}^{w_c}$ associated with the root system $\Phi_c(x)$. Hence there is an element in the Weyl group of $\Phi_c(x)$ on $i\mathfrak{t}^{w_c}$ which carries $\tilde{s}$

to $\tilde{s}'$. By Proposition 7.2.5, the Weyl group of $\Phi_c(x)$ is the Weyl group of $Z(x, y_0)$. Thus, there exists a $g \in N_T(Z(x, y_0))$ conjugating $\tilde{s}$ to $\tilde{s}'$, and hence (x, sy_0) and $(x, s'y_0)$ are conjugate c-pairs in G.

This allows us to assume further that $y = y'$. By Proposition 8.1.2, $Z^0(x, y) = S^{w_c}$. By Corollary 7.5.3, the group $\pi_0(Z(x, y))$ is cyclic of order dividing $g_{\overline{a}}$, and $[z]$ is an element of this group of order exactly $g_{\overline{a}}$. Hence $[z]$ generates $\pi_0(Z(x, y))$. Thus there is a unique integer ℓ with $1 \leq \ell < g_{\overline{a}}$ such that z^ℓ and z' are in the same component of $Z(x, y)$. Since z' acts on S^{w_c} with isolated fixed points, this implies that z^ℓ and z' are conjugate in $Z(x, y)$ and hence that (x, y, z^ℓ) and (x, y, z') are conjugate in G. $\square$

8.1.5. Simple groups containing c-triples of rank zero. It follows from the above that a simple group G has a c-triple of rank zero and order k if and only if exactly one of the integers $g_{\overline{a}}$ is divisible by k. Examining the quotient diagrams gives the following list of all the possibilities for $c \neq 1$:

(1) $G = A_n$, c a generator of $\mathcal{C}G$ and $k | n + 1$.
(2) $G = C_2$, c the non-trivial element of $\mathcal{C}G$ and $k = 2$.
(3) $G = D_6$, c an exotic element of the center and $k = 4$.
(4) $G = E_6$, c a generator of $\mathcal{C}G$ and $k = 2$ or 6.
(5) $G = E_7$, c the non-trivial element of $\mathcal{C}G$ and $k = 3$ or 6.

8.1.6. Action of the automorphism group of G on rank zero c-triples.

LEMMA 8.1.8. *Let (x, y) be a c-pair in normal form in G. Let $g \in G$ and suppose that $i_g(x) = \zeta x$ for some $\zeta \in \mathcal{C}G$. Let $\sigma \colon G \to G$ be an automorphism normalizing S^{w_c} such that $\sigma(x) = \zeta_1 x$ and $\sigma(y) = \zeta_2 y$, where $\zeta_1, \zeta_2 \in \mathcal{C}G$. Then both i_g and σ induce the identity on* $\mathrm{Tor}\,((Q^\vee/Q^\vee(x))_{w_c})$. *Moreover, σ induces the identity on $\pi_0(Z(x, y))$.*

PROOF. Since T is a maximal torus for $Z(x)$, there is an inner automorphism i_h, $h \in Z(x)$, such that $i_h \circ i_g$ normalizes T. Since $Z(x)$ is connected, the inner automorphism i_h of $Z(x)$ induces the identity on its fundamental group. Thus, without loss of generality, we can assume that g normalizes T. Let $\tilde{x} \in i\mathfrak{t}$ be an element in the alcove A such that $\exp(2\pi i\tilde{x}) = x$. Then $g\tilde{x}g^{-1} = \tilde{x} + \tilde{\zeta}$ for some element $\tilde{\zeta} \in i\mathfrak{t}$ with $\exp(2\pi i\tilde{\zeta}) = \zeta$. Since $\zeta \in \mathcal{C}G$, the affine automorphism $t \mapsto \varphi(t) = gtg^{-1} + \tilde{\zeta}$ of $i\mathfrak{t}$ normalizes the alcove decomposition associated with the root system of G. Since $\varphi(\tilde{x}) = \tilde{x}$, $\varphi(A) = A'$ is an alcove containing $\tilde{x}$. Hence there is an element μ of the affine Weyl group of G which fixes $\tilde{x}$ and sends A' to A. The composition $\mu \circ \varphi$ then normalizes A. Since the translational part of this affine transformation is congruent to $\tilde{\zeta}$ modulo the coroot lattice $Q^\vee$, we see by Lemma 3.2.1 that this composition is $\nu_A(\zeta)$, which we denote by w_ζ. On the other hand, since μ fixes $\tilde{x}$, the Weyl part w of μ is an element of the Weyl group of $Z(x)$. Since $w \circ i_g|\mathfrak{t} = w_\zeta$, multiplying g by an element of $N_{Z(x)}(T)$ allows us to assume, without loss of generality that $i_g|\mathfrak{t} = w_\zeta$.

As noted in Section 3.2, we have $g_{w_\zeta(a)} = g_a$ for all $a \in \tilde{\Delta}$. Since w_ζ and w_c commute, we also have $g_{w_\zeta(\overline{a})} = g_{\overline{a}}$ for all $\overline{a} \in \tilde{\Delta}_c$. Now we see directly from the expression for a generator of $\mathrm{Tor}\,((Q^\vee/Q^\vee(x))_{w_c})$ given in Lemma 7.5.2, that w_ζ acts trivially on $\mathrm{Tor}\,((Q^\vee/Q^\vee(x))_{w_c})$. This completes the proof of the first part of the lemma.

Since $S^{w_c} \cdot x$ contains a generic element of G, σ normalizes T. The equation $\sigma(y) = \zeta_2 y$ says that $\sigma|\mathfrak{t}$ commutes with $i_y = w_c$. There is a Weyl element $w \in W(G)$ commuting with w_c such that $w(x) = \zeta_1^{-1}x$ and such that $w \circ \sigma(A) = A$. Since $w \circ \sigma(A) = A$, it follows that $g_{w\circ\sigma(a)} = g_a$ for all $a \in \widetilde{\Delta}$. Since $w \circ \sigma$ commutes with w_c, it is also the case that $g_{w\circ\sigma(\overline{a})} = g_{\overline{a}}$ for all $\overline{a} \in \widetilde{\Delta}_c$. Thus, $w \circ \sigma$ acts trivially on $\mathrm{Tor}\left((Q^\vee/Q^\vee(x))_{w_c}\right)$. The first part of the lemma implies that w acts trivially on this group. Thus, it follows that σ acts trivially on $\mathrm{Tor}\left((Q^\vee/Q^\vee(x))_{w_c}\right)$. The final statement follows since the inclusion $\pi_0(Z(x,y)) \to \mathrm{Tor}\left((Q^\vee/Q^\vee(x))_{w_c}\right)$ of Lemma 7.3.2 is equivariant with respect to the automorphism σ. □

LEMMA 8.1.9. *If (x,y,z) is a c-triple of rank zero in G and if $\sigma\colon G \to G$ is an automorphism of G fixing c, then $\sigma(x,y,z)$ is conjugate to (x,y,z).*

PROOF. We can assume that (x,y) is a c-pair in normal form. Since $\sigma(c) = c$, the triple $(\sigma(x), \sigma(y), \sigma(z))$ is also a c-triple, clearly of rank zero. By Proposition 8.1.7, composing σ with an inner automorphism allows us to assume that $\sigma(x,y,z) = (x,y,z^\ell)$ for some integer ℓ. By Lemma 8.1.2, since (x,y,z) is of rank zero, $Z^0(x,y) = S^{w_c}$. Hence, $\sigma(S^{w_c}) = S^{w_c}$. Applying Lemma 8.1.8 we see that σ acts trivially on $\mathrm{Tor}\left((Q^\vee/Q^\vee(x))_{w_c}\right)$. Hence, by Lemma 7.3.2, it follows that $z' = \sigma(z)$ and z are in the same component of $Z(x,y)$, and hence that $\sigma(x,y,z)$ is conjugate to (x,y,z). □

8.1.7. Action of $\mathcal{C}G$ on the space of rank zero c-triples. We shall first consider the action on the first component.

LEMMA 8.1.10. *Let $\gamma \in \mathcal{C}G$. Then $\gamma \cdot (x,y,z) = (\gamma x, y, z)$ is the trivial action on the set of conjugacy classes of c-triples of rank zero in G.*

PROOF. Note that $Z(\gamma x, y) = Z(x,y)$. It follows that the order of $(\gamma x, y, z)$ is equal to that of (x,y,z), and it clearly has rank zero. We can assume that (x,y) is in normal form with respect to A. By Proposition 8.1.7 there is $g \in G$ which conjugates $(\gamma x, y)$ to (x,y). Since $i_g(x) = \gamma^{-1}x$ and $i_g(y) = y$, i_g induces an automorphism of $Z(x,y)$. By Lemma 8.1.8, the induced action of i_g on $\pi_0(Z(x,y))$ is trivial. In particular, $i_g(z)$ and z lie in the same connected component of $Z(x,y)$. Hence they are conjugate in $Z(x,y)$. It follows that $(\gamma x, y, z)$ and (x,y,z) are conjugate in G. □

LEMMA 8.1.11. *For any $\gamma \in \mathcal{C}G$ the action $\gamma \cdot (x,y,z) = (x, \gamma y, z)$ is trivial on the space of conjugacy classes of c-triples of rank zero in G.*

PROOF. By symmetry between x and y (at the expense of replacing c by c^{-1}), the result in this case follows from the previous one. □

Lastly, we must consider the action $\gamma \cdot (x,y,z) = (x,y,\gamma z)$.

LEMMA 8.1.12. *For $\gamma \in \mathcal{C}$ the action of $\mathcal{C}G$ defined by $\gamma \cdot (x,y,z) = (x,y,\gamma z)$ induces an action of $\mathcal{C}G$ on the set of conjugacy classes of rank zero c-triples in G. The stabilizer $\mathcal{K}$ of a conjugacy class of a c-triple of rank zero is $\mathcal{K} = \mathcal{C}G \cap S^{w_c}$. The order of each orbit, or equivalently, the index of $\mathcal{K}$ in $\mathcal{C}G$ is $\gcd(g_{\overline{a}}) = n_0$.*

PROOF. Let (x,y,z) be a c-triple of rank zero such that (x,y) is a c-pair in normal form. Then $Z^0(x,y) = S^{w_c}$. For any $\gamma \in \mathcal{C}G$, γz and z are in the same component of $Z(x,y)$ if and only if $\gamma \in S^{w_c}$. This proves that $\mathcal{K} = \mathcal{C}G \cap S^{w_c}$.

Lastly, we show that the order of $\mathcal{C}G/(S^{w_c}\cap \mathcal{C}G)$ is equal to n_0. By Corollary 7.5.3, n_0 is the number of components of T^{w_c}. Since $\mathcal{C}G\subseteq T^{w_c}$ and since S^{w_c} is the component of the identity of T^{w_c}, this shows that the order of $\mathcal{C}G/(S^{w_c}\cap \mathcal{C}G)$ divides n_0.

To complete the proof, we need to show that the inclusion $\mathcal{C}G\to T^{w_c}$ is surjective on the level of components. We state this as a separate lemma:

LEMMA 8.1.13. *Let G be a simple group and let $c\in \mathcal{C}G$. The inclusion $\mathcal{C}G\to T^{w_c}$ induces a surjection $\mathcal{C}G\to \pi_0(T^{w_c})$.*

PROOF. Applying cohomology to the w_c-actions on the exact sequence

$$0\to Q^\vee\to P^\vee\to \mathcal{C}G\to 0$$

and considering the torsion subgroups, we have an exact sequence

$$\mathcal{C}G\to \mathrm{Tor}(Q^\vee_{w_c})\to \mathrm{Tor}(P^\vee_{w_c})\to \mathcal{C}G.$$

Thus, it suffices to show that $\mathrm{Tor}(P^\vee_{w_c})\to \mathcal{C}G$ is injective.

Set $h_{\overline{a}}=n_{\overline{a}}h_a$. There is the exact sequence

$$0\to \mathbf{Z}(\sum_{a\in\widetilde{\Delta}} h_a a)\to \bigoplus_{a\in\widetilde{\Delta}}\mathbf{Z}(a)\to Q\to 0,$$

where $Q=Q(\Phi)$ is the root lattice of G. Dualizing gives

$$0\to P^\vee\to \bigoplus_{a\in\widetilde{\Delta}}\mathbf{Z}a^*\to \mathbf{Z}\to 0,$$

where the second map is obtained by sending $\sum_a r_a a^*$ to $\sum_a r_a h_a$. (Here the a^* are the dual basis to the basis $\{a : a\in\widetilde{\Delta}\}$ of the free $\mathbf{Z}$-module $\bigoplus_{a\in\widetilde{\Delta}}\mathbf{Z}(a)$.) Hence, the group $\mathrm{Tor}(P^\vee_{w_c})$ is a cyclic group of order $\gcd(h_{\overline{a}})$. Clearly, if $\overline{a}$ contains the extended root, $n_{\overline{a}}$ is the order of c and $h_{\tilde{a}}=1$ so that $h_{\overline{a}}$ is the order of c. Thus the order of $\mathrm{Tor}(P^\vee_{w_c})$ divides the order of c. On the other hand, a generator for $\mathrm{Tor}(P^\vee_{w_c})$ is represented by $a^*-w_c(a^*)\in \mathcal{C}G$ for any $a\in\widetilde{\Delta}$ for which $h_a=1$. Choose a to be the root mapped by w_c to $\tilde{a}$. Then the corresponding element of $\mathrm{Tor}(P^\vee_{w_c})$ is represented by $\varpi^\vee_a\in P^\vee$, and its image in $\mathcal{C}G$ is therefore equal to c^{-1}. Since the order of $\mathrm{Tor}(P^\vee_{w_c})$ is at most that of c, the map $\mathrm{Tor}(P^\vee_{w_c})\to \mathcal{C}G$ is injective, and in fact an isomorphism. This proves the lemma. □

8.2. The maximal torus of a c-triple of order k

Fix a non-trivial element $c\in\mathcal{C}G$. Let (x,y,z) be a c-triple in G of order k. We let S be a maximal torus for $Z(x,y,z)$ and let $\mathfrak{s}$ be its Lie algebra. As usual, $L=DZ(S)$. By Theorem 2.3.1 there is a c-triple $(x_0,y_0,z_0)\in L$ of rank zero and $s_1,s_2,s_3\in S$ such that $(x,y,z)=(s_1x_0,s_2y_0,s_3z_0)$.

8.2.1. Determination of $\mathfrak{s}$.

Our goal is to describe the torus S, or equivalently $\mathfrak{s}$. We begin with the following definition.

DEFINITION 8.2.1. Let $\overline{\mathbf{g}}\colon \widetilde{\Delta}_c\to \mathbf{N}$ be the function defined by $\overline{\mathbf{g}}(\overline{a})=g_{\overline{a}}$. For each integer $k\geq 1$ dividing at least one of the integers $g_{\overline{a}}$, for $\overline{a}\in\widetilde{\Delta}_c$, we define $f^c(k)$ to be the maximal face of A^c with the property that every root $\overline{a}\in\widetilde{\Delta}_c$ for which $k\nmid g_{\overline{a}}$ takes an integral value on $f^c(k)$. Let $\mathfrak{t}^{w_c}(\overline{\mathbf{g}},k)$ be the linear subspace of $\mathfrak{t}^{w_c}$ parallel to $f^c(k)$ (by which we mean that $i\mathfrak{t}^{w_c}$ is parallel to $f^c(k)$ in $i\mathfrak{t}$) and let $S^{w_c}(\overline{\mathbf{g}},k)$ be the torus whose Lie algebra is $\mathfrak{t}^{w_c}(\overline{\mathbf{g}},k)$.

We can then state the main result as follows:

PROPOSITION 8.2.2. *Let (x,y,z) be a c-triple of order k and let S be a maximal torus of $Z(x,y,z)$. Then k divides at least one of the integers $g_{\overline{a}}$ and S is conjugate to $S^{w_c}(\mathbf{\overline{g}},k)$. The element x is conjugate to a point of the form $\exp(2\pi i\tilde{x})$ for some $\tilde{x}\in f^c(k)$. In particular, the dimension of S is equal to one less than the number of $\overline{a}\in\widetilde{\Delta}$ for which k divides $g_{\overline{a}}$.*

First let us show that $\mathfrak{s}$ is conjugate to a linear subspace parallel to some face of A^c:

LEMMA 8.2.3. *Let (x,y,z) be a c-triple such that (x,y) is a c-pair in normal form. Suppose that $x'\in S\cdot x$ is generic in the sense that any root of G equal to one on x' is equal to one on $S\cdot x$ and suppose that $x'=\exp(2\pi i\tilde{x}')$ for some $\tilde{x}'$ contained in A^c. Let f be the face of A^c containing $\tilde{x}'$ in its interior. Then, up to conjugation by an element in $Z(x,y)$, $\mathfrak{s}$ is the linear subspace of $\mathfrak{t}^{w_c}$ parallel to f.*

PROOF. Since $x'\in S\cdot x$ is generic, it follows that $\dim\mathfrak{s}\le\dim f$. On the other hand, for every $\tilde{x}''$ in the interior of f, the triple $(\exp(2\pi i\tilde{x}''),y,z)$ is a c-triple with the same centralizer as the triple (x,y,z). This means that $f-\tilde{x}'$ is contained in $\mathfrak{s}$. Comparing this with the dimension estimate shows that $\mathfrak{s}$ is the linear subspace parallel to f. □

LEMMA 8.2.4. *Let (x,y,z) be a c-triple of order k. Then k divides at least one of the $g_{\overline{a}}$ and x is conjugate to $\exp(2\pi i\tilde{x})$ for some $\tilde{x}\in f^c(k)$. Moreover, the Lie algebra $\mathfrak{s}$ of S is conjugate to the linear subspace of $\mathfrak{t}^{w_c}$ parallel to a face of A^c contained in $f^c(k)$.*

PROOF. After conjugation, we may assume that (x,y) is a c-pair in normal form. Since (x,y,z) is of order k, the order of $\pi_0(Z(x,y))$ is divisible by k, and hence $(Q^\vee/Q^\vee(x))_{w_c}$ has order divisible by k. By Lemma 7.5.2, this means $n=\gcd_{a\in\widetilde{\Delta}-\widetilde{I}(x)}\{g_{\overline{a}}\}$ is divisible by k, and in particular $k|g_{\overline{a}}$ for some a. Moreover $\{\overline{a}\mid k\nmid g_{\overline{a}}\}\subseteq\widetilde{I}(x)$. Thus, x is the image under $\exp(2\pi i\cdot)$ of a point $\tilde{x}$ contained in $f^c(k)$.

Take $x'\in S\cdot x$ to be a generic element with the property that $x=\exp(2\pi i\tilde{x}')$ for some $\tilde{x}'\in A^c$. Let f be the face of A^c containing $\tilde{x}'$ in its interior. By the first part of this lemma, f is a face of $f^c(k)$. According to the previous lemma $\mathfrak{s}$ is parallel to f. □

We turn now to the proof of Proposition 8.2.2. As usual, let $L=DZ(S)$ and let (x_0,y_0,z_0) be a rank zero c-triple in L such that $(x,y,z)=(s_1x_0,s_2y_0,s_3z_0)$ for elements $s_1,s_2,s_3\in S$. Let $\mathfrak{t}_L=\mathrm{Lie}(L)\cap\mathfrak{t}$; it is the Lie algebra of a maximal torus for L. Let $Q_L^\vee=Q^\vee\cap i\mathfrak{t}_L$. Of course, $c\in L$ and the action of w_c on $\mathfrak{t}$ normalizes $\mathfrak{t}_L$. The group L is simply connected and semi-simple, but may not be simple. Let $\prod_i L_i$ be its decomposition into simple factors and let Φ_{L_i} be the corresponding root system with respect to $\mathrm{Lie}(L_i)\cap\mathfrak{t}$. We decompose $c\in L$ as $\prod_i c_i$. Write $x_0=\prod_i x_{0,i}$, and similarly for y_0 and z_0. Then $(x_{0,i},y_{0,i},z_{0,i})$ is a rank zero c_i-triple in L_i for every i. Of course, $\mathrm{Tor}\left((Q_L^\vee/Q_L^\vee(x_0))_{w_c}\right)=\bigoplus_i\mathrm{Tor}\left((Q_{L_i}^\vee/Q_{L_i}^\vee(x_{0,i}))_{w_{c_i}}\right)$. By Lemma 7.5.2 applied to the $(x_{0,i},y_{0,i})$, the groups $\mathrm{Tor}\left((Q_{L_i}^\vee/Q_{L_i}^\vee(x_{0,i}))_{w_{c_i}}\right)$ are cyclic. Let $\mu_i\in\mathrm{Tor}\left((Q_{L_i}^\vee/Q_{L_i}^\vee(x_{0,i}))_{w_{c_i}}\right)$ be the element $\delta([z_{0,i}])$, where $[z_{0,i}]\in\pi_0(Z(x_{0,i},y_{0,i}))$ is the class of $z_{0,i}$ and δ is the homomorphism of Lemma 7.3.2.

By Lemma 7.3.5, there is is an element $\tilde{\mu}_i \in (Q^\vee_{L_i}(x_{0,i})_{w_{c_i}}) \otimes \mathbf{Q}$ whose image in $\mathrm{Tor}\left((Q^\vee_{L_i}/Q^\vee_{L_i}(x_{0,i}))_{w_{c_i}}\right)/\mathrm{Tor}((Q^\vee_{L_i})_{w_{c_i}})$ is the image of μ_i under the quotient map. According to Proposition 8.1.4, the coefficients of $\tilde{\mu}_i$ are all non-integral when written as a linear combination of any set of simple coroots for the root system $\Phi^{\mathrm{proj}}_{Z_{L_i}(x_{0,i})}(w_c)$. Clearly

$$\delta([z_0]) = \sum_i \mu_i \in \bigoplus_i (Q^\vee_{L_i}/Q^\vee_{L_i}(x_{0,i}))_{w_c} = (Q^\vee_L/Q^\vee_L(x_0))_{w_c}.$$

Let $\mu = \sum_i \mu_i$ and let $\tilde{\mu} = \sum_i \tilde{\mu}_i \in (Q^\vee_L(x_0)_{w_c}) \otimes \mathbf{Q}$. Then $\tilde{\mu}$ projects to the image of μ in $\mathrm{Tor}\left((Q^\vee_L/Q^\vee_L(x_0))_{w_c}\right)/\mathrm{Tor}((Q^\vee_L)_{w_c})$. For every set of simple roots for $\Phi^{\mathrm{proj}}_{Z_L(x_0)}(w_c)$, if we write $\tilde{\mu}$ as a linear combination of the corresponding coroots, then no coefficient is integral.

Fix a generic element $x' \in S \cdot x_0$. Then $Z_G(x') = S \cdot Z_L(x_0) \subseteq Z_G(x)$ and so $\Phi(x') = \Phi_L(x_0)$. Hence $Q^\vee_L(x_0) = Q^\vee(x')$. Moreover $Z(x', y) \subseteq Z(x, y)$. Furthermore, $Z_G(x', y_0) = Z_G(x', y) = S \cdot Z_L(x_0, y_0)$. Clearly, $z \in Z(x', y)$. Since the order of $[z] \in \pi_0(Z(x, y))$ is k, it follows from Lemma 7.6.2 that the order of $[z] \in \pi_0(Z(x', y))$ is also k. Thus the element $\mu \in (Q^\vee/Q^\vee(x'))_{w_c}$ is of order k. Since $\Phi(x') = \Phi_L(x_0)$, $\Phi_c(x') = \Phi^{\mathrm{proj}}_{Z(x')}(w_c) = \Phi^{\mathrm{proj}}_{Z_L(x_0)}(w_c)$. Thus, expressing $\tilde{\mu}$ as a linear combination of $\pi(a^\vee)$ for $\overline{a} \in \widetilde{I}_c(x')$, no coefficient of $\tilde{\mu}$ is integral. It follows by Lemma 8.1.5 that $k \nmid g_{\overline{a}}$ for any $\overline{a} \in \widetilde{I}_c(x')$. Hence, for a generic element $x' \in S \cdot x_0$ the subset $\widetilde{I}_c(x')$ is contained in the subset of $\overline{a} \in \widetilde{\Delta}_c$ for which $k \nmid g_{\overline{a}}$. This together with Lemma 8.2.4 shows that the generic $x' \in S \cdot x_0$ is conjugate to a point for the form $\exp(2\pi i\tilde{x})$ for $\tilde{x}$ in the interior of the face $f^c(k)$ and hence that $\mathfrak{s}$ is exactly the linear space parallel to $f^c(k)$. This completes the proof of Proposition 8.2.2. □

The following is a corollary of the proof:

COROLLARY 8.2.5. *Let (x, y, z) be a c-triple and let (x_0, y_0, z_0) be any rank zero c-triple in L such that $(x, y, z) = (s_1x_0, s_2y_0, s_3z_0)$, where the $s_i \in S$. Then the order of (x, y, z) is the order of $[z_0] \in \pi_0(Z_L(x_0, y_0))/\pi_0(S \cap Z_L(x_0, y_0))$, and hence the order is constant on connected components of $\mathcal{T}_G(c)$. Finally, the order of the c-triple (x, y, z) of G divides the order of the c-triple (x_0, y_0, z_0) in L.*

REMARK 8.2.6. The order of (x, y, z) in G is not always the order of (x_0, y_0, z_0) in L.

By Proposition 8.2.2, if G has a c-triple of order k, then k divides at least one of the $g_{\overline{a}}$. Just as in the case of commuting triples, there is a converse to this statement:

PROPOSITION 8.2.7. *Let k be a positive integer. There is a c-triple of order k in G if and only if k divides at least one of the $g_{\overline{a}}$.*

PROOF. The "only if" direction follows from Proposition 8.2.2. Conversely, suppose that k divides at least one of the $g_{\overline{a}}$. If the w_c-action of $\widetilde{\Delta}$ has an exceptional orbit, then all of the $g_{\overline{a}}$ are equal. The result follows easily in this case from Proposition 7.5.5. Otherwise, choose $x = \exp(2\pi i\tilde{x})$ with $\tilde{x} \in f^c(k)$. Then k divides the order of $\mathrm{Tor}\left((Q^\vee/Q^\vee(x))_{w_c}\right)$. It then follows by Corollary 7.3.11 that there is a choice of y such that (x, y) is a c-pair and such that k divides the order of $\pi_0(Z(x, y))$. Choose a $z \in Z(x, y)$ mapping to an element of order k in $\pi_0(Z(x, y))$. Then (x, y, z) is a c-triple in G of order k. □

8.2.2. A normal form for c-triples of order k. Let $\mathbf{x}$ be a c-triple and let S, L be as above. Write $\mathbf{x} = (s_1x_0, s_2y_0, s_3z_0)$ where the $s_i \in S$ and $(x_0, y_0, z_0) \in L$. Let $L = \prod_i L_i$ be the decomposition of L as a product of simple groups and let $c = c_1 \cdots c_r$ be the corresponding decomposition of c. We can write x_0 as a product of elements $x_{0,i}$, and similarly for y_0, z_0. An alcove for L is a product of alcoves for the L_i. Likewise, the root system $\Phi_c(x_0)$ on $i(\mathfrak{t}_L)^{w_c}$ is a product of the root systems $\Phi_{c_i}(x_{0,i})$ on $i(\mathfrak{t}_{L_i})^{w_{c_i}}$. If (x_0, y_0) is a c-pair in L, we say that (x_0, y_0) is in normal form for the product of the B_i if each c_i-pair $(x_{0,i}, y_{0,i})$ is in normal form for B_i.

With this notation, we have the following:

PROPOSITION 8.2.8. *Suppose that k divides at least one of the $g_{\overline{a}}$ for $\overline{a} \in \widetilde{\Delta}_c$. Let $S \subseteq T^{w_c}$ be the torus whose Lie algebra $\mathfrak{s}$ is parallel to the face $f^c(k)$ of A^c. Let $L = DZ(S)$, and let $\mathfrak{t}_L = \mathfrak{t} \cap \mathrm{Lie}(L) = \mathfrak{s}^\perp$. Let $\tilde{x}_0$ be the unique point such that $(i\mathfrak{s} + f^c(k)) \cap (i\mathfrak{t}_L) = \{\tilde{x}_0\}$, and let $x_0 = \exp(2\pi i\tilde{x}_0) \in L$. Finally let $y_0 = sy_1$, where y_1 is such that (x_0, y_1) is a special c-pair in normal form and $s = \exp(2\pi ib)$, where b is the product of the barycenters of the alcoves B_i for the root systems $\Phi_{c_i}(x_{0,i})$ in the simple factors of L. If $\mathbf{x}$ is a c-triple of order k in G, then there are elements s_1, s_2, s_3 in S and $z_0 \in L$ such that (s_1x_0, s_2y_0, s_3z_0) is a c-triple conjugate to $\mathbf{x}$.*

PROOF. Write $\mathbf{x} = (x, y, z)$. Let S be a maximal torus for $Z(\mathbf{x})$. By Proposition 8.2.2, possibly after conjugating $\mathbf{x}$, we can assume that S is the torus whose Lie algebra is parallel to $f^c(k)$ and that $x = \exp(2\pi i\tilde{x})$ for some $\tilde{x} \in f^c(k)$. Thus $x = s_1x_0$ for some $s_1 \in S$. There are $s_2, s_3 \in S$ such that $y = s_2y_0$ and $z = s_3z_0$, where $y_0, z_0 \in L$. After a further conjugation in L, we can assume that (x_0, y_0) is a c-pair in normal form for L. Note that (x_0, y_0, z_0) has rank zero in L. It then follows from Proposition 8.1.4, applied to the simple factors of L, that y_0 is as described in the statement of the proposition. □

8.2.3. More on the group L. Let $\mathbf{x}$ be a c-triple in G, let S be a maximal torus for $Z(\mathbf{x})$, and let $L = DZ(S)$.

PROPOSITION 8.2.9. *Let $n_0 = \gcd\{g_{\overline{a}}\}$. There is at most one simple factor of L which is not of type A. Moreover, the following are equivalent:*

(1) *The order k of the c-triple (x, y, z) divides n_0.*
(2) *S is conjugate to S^{w_c}.*
(3) *L is conjugate to $L_c = DZ(S^{w_c})$.*
(4) *Every simple factor of L is of type A.*

PROOF. As in the case of commuting triples, since the Dynkin diagram of L is a proper subdiagram of the Dynkin diagram of G, L can have at most one simple factor which is not of A-type. This proves the first statement.

To prove the equivalences of the proposition, note that, by Proposition 8.2.2, $S = S^{w_c}$ if and only if k divides $g_{\overline{a}}$ for every $\overline{a}$. Thus (1) is equivalent to (2). If $S = S^{w_c}$, then by definition $L = DZ(S^{w_c}) = L_c$. Thus (2) implies (3). Conversely, if $L = L_c$, then S is the torus associated to the intersections of the kernels of the roots of L_c, and thus $S = S^{w_c}$.

Since, by Proposition 3.5.1, all simple factors of L_c are of type A, (3) implies (4). Finally, we show that (4) implies (3). Let (x_0, y_0, z_0) be a c-triple of rank zero in L such that $(x, y, z) = (s_1x_0, s_2y_0, s_3z_0)$, where the $s_i \in S$. Write $x_0 = \prod_i x_{0,i}$, where the $x_{0,i}$ lie in the simple factors L_i of L, and similarly for y_0, z_0. Since L_i is of type A, $(x_{0,i}, y_{0,i})$ is a rank zero c_i-pair in L_i. It follows that L contains the rank zero c-pair (x_0, y_0). Since $c \in L$, it follows by Lemma 3.4.1 that $L_c \subseteq L$. Since L contains a rank zero c-pair, it follows that $L \subseteq L_c$. Thus $L = L_c$. □

We can say more about L in case there is a simple factor not of type A:

PROPOSITION 8.2.10. *Suppose that L contains a simple factor, not of type A, and write $L = L_0 \times \prod_{i=1}^{s} L_i$, where L_0 is simple and not of type A, and the L_i are simple and of type A for $i \geq 1$. Let $c = \prod_{i=0}^{s} c_i$ be the corresponding decomposition of c. Finally let L_{0,c_0} be the subgroup of L_0 associated to the element $c_0 \in \mathcal{C}L_0$ as in Section 3.4. Then $L_c = L_{0,c_0} \times \prod_{i=1}^{s} L_i$.*

PROOF. Since $c_0 \in L_{0,c_0}$, we have $c \in L_{0,c_0} \times \prod_{i=1}^{s} L_i$. Moreover, $L_{0,c_0} \times \prod_{i=1}^{s} L_i$ is a product of groups of A-type, and c projects to a generator of the center of each factor. The result now follows from Lemma 3.5.2. □

8.3. The number of components

Our goal here is to describe the number of components of $\mathcal{T}_G(c)$ of order k and to identify each such component explicitly. We will postpone the explicit determination of the Weyl group $W(S, G)$ to Chapter 9, however.

Suppose that k divides at least one of the $g_{\overline{a}}$ for $\overline{a} \in \widetilde{\Delta}_c$. We keep the notation of Proposition 8.2.8. In particular, let $\tilde{x}_0$ be the unique point such that $(i\mathfrak{s}+f^c(k)) \cap (i\mathfrak{t}_L) = \{\tilde{x}_0\}$, and let $x_0 = \exp(2\pi i \tilde{x}_0) \in L$, and let $y_0 = sy_1$, where y_1 is such that (x_0, y_1) is a special c-pair in normal form and s is the exponential of the products of the barycenters of the alcoves B_i for the root systems $\Phi_{c_i}(x_{0,i})$ in the simple factors of L. Then (x_0, y_0) is a c-pair in L.

PROPOSITION 8.3.1. *Let $\mathbf{x}$ be a c-triple of order k in G. Then $\mathbf{x}$ is conjugate to a c-triple of the form (s_1x_0, s_2y_0, s_3z_0) where x_0, y_0 are as above, $z_0 \in L$ and $s_1, s_2, s_3 \in S$. Denote by $\psi(\mathbf{x})$ the class $[z_0] \in \pi_0(Z_G(x_0, y_0))$.*

(1) *$\psi(\mathbf{x})$ is well-defined and depends only on the conjugacy class of $\mathbf{x}$.*
(2) *ψ is constant on the components of $\mathcal{T}_G(c)$ of order k in G.*
(3) *$\psi(\mathbf{x})$ is of order k.*
(4) *ψ induces a bijection from the set of components of $\mathcal{T}_G(c)$ of order k in G to the set of elements of order k in $\pi_0(Z_G(x_0, y_0))$.*

PROOF. By Proposition 8.2.8, $\mathbf{x}$ is conjugate to (s_1x_0, s_2y_0, s_3z_0) as claimed. To prove Part (1), it is clearly sufficient to show the following: suppose that (s_1x_0, s_2y_0, s_3z_0) and $(t_1x_0, t_2y_0, t_3z_0')$ are two c-triples which are conjugate in G. Then $[z_0] = [z_0']$ in $\pi_0(Z_G(x_0, y_0))$. By Theorem 2.3.1, there is a $g \in N_G(S)$ such that

$$i_g(x_0, y_0, z_0) = (u_1x_0, u_1y_0, u_3z_0'),$$

where $u_i \in S \cap L$. Moreover we can assume that $g \in N_G(T)$. Clearly $[z_0'] = [i_g(z_0)]$, and we must show that $[z_0] = [i_g(z_0)]$.

First note that, since $[g, y_0] \in \mathcal{C}L \subseteq T$, i_g commutes with w_c, and thus i_g acts on T^{w_c} and hence on $\pi_0(T^{w_c})$. In particular i_g acts on L_c. By Lemma 8.1.13, the center $\mathcal{C}G$ surjects onto $\pi_0(T^{w_c})$, and so the induced action of i_g on $\pi_0(T^{w_c})$ is trivial.

Since $S \cap L \subseteq \mathcal{C}L$, the inner automorphism i_g defines an automorphism of $Z_L(x_0, y_0)$. Moreover, $[z_0]$ is the image of the element $[z_0] \in \pi_0(Z_L(x_0, y_0))$ under the natural homomorphism $\rho\colon \pi_0(Z_L(x_0,y_0)) \to \pi_0(Z_G(x_0,y_0))$. We must show that $\rho\circ i_g=\rho$. We may write $L\cong L_0\times\prod_{i\geq 1}L_i$, where the L_i are simple groups of type A and L_0 is either trivial or a simple group not of type A. Clearly i_g preserves the factors L_0 and $\prod_{i\geq 1}L_i$. There is a corresponding direct sum decomposition

$$\pi_0(Z_L(x_0,y_0))=\pi_0(Z_{L_0}(x_{0,0},y_{0,0}))\oplus\bigoplus_{i\geq 1}\pi_0(Z_{L_i}(x_{0,i},y_{0,i})),$$

where the $x_{0,i}, y_{0,i}$ are the projections of x_0, y_0 to the factor L_i. We analyze the action on each of these factors separately.

The group $\prod_{i\geq 1}L_i$ is a subgroup of L_c. Thus the map

$$\bigoplus_{i\geq 1}\pi_0(Z_{L_i}(x_{0,i},y_{0,i}))\to\pi_0(Z_G(x_0,y_0))$$

factors through $\pi_0(Z_{L_c}(x_0,y_0))$. But $\pi_0(Z_{L_c}(x_0,y_0))=Z_{L_c}(x_0,y_0)=\mathcal{C}L_c$. Since $\mathcal{C}L_c\subseteq T^{w_c}$, the map $\pi_0(Z_{L_c}(x_0,y_0))\to\pi_0(Z_G(x_0,y_0))$ factors through $\pi_0(T^{w_c})$. But as we have seen, i_g acts trivially on $\pi_0(T^{w_c})$. Thus, for any element $\xi'\in\bigoplus_{i\geq 1}\pi_0(Z_{L_i}(x_{0,i},y_{0,i}))$, we have $\rho(i_g(\xi'))=\rho(\xi')$.

Now consider the action of i_g on $\pi_0(Z_{L_0}(x_{0,0},y_{0,0}))$. Since i_g defines an automorphism of Z_{L_0} which sends $(x_{0,0},y_{0,0})$ to an element of the form $(v_1x_{0,0},v_2y_{0,0})$, where the $v_i\in\mathcal{C}L_0$, Lemma 8.1.8 implies that the action of i_g on $\pi_0(Z_{L_0}(x_{0,0},y_{0,0}))$ is trivial. Thus, for $\xi_0\in\pi_0(Z_{L_0}(x_{0,0},y_{0,0}))$, $\rho(i_g(\xi_0))=\rho(\xi_0)$. It follows that $\rho\circ i_g=\rho$, and finally that $[z_0]=[i_g(z_0)]$ as claimed.

To see (2), if the conjugacy class of $\mathbf{x}'=(x',y',z')$ is in the same component of $\mathcal{T}_G(c)$ as that of $\mathbf{x}$, then after conjugation we can assume that $(x',y',z')=(t_1x_0,t_2y_0,t_3z_0)$ for $t_i\in S$. Then by the definition of ψ and the fact that it is well-defined, we see that $\psi(\mathbf{x}')=[z_0]=\psi(\mathbf{x})$.

To see (3), set $\mathbf{x}_0=(x_0,y_0,z_0)$. Then $\psi(\mathbf{x})=\psi(\mathbf{x}_0)=[z_0]$. On the other hand, $\mathbf{x}_0$ is a c-triple in the same component of $\mathcal{T}_G(c)$ as $\mathbf{x}$, so by Corollary 8.2.5, the order of $\mathbf{x}_0$ is k. By definition, this is the order of $[z_0]=\psi(\mathbf{x}_0)=\psi(\mathbf{x})$ in $\pi_0(Z_G(x_0,y_0))$.

Finally, we prove (4). Suppose that $\psi(\mathbf{x}_1)=\psi(\mathbf{x}_2)$. Then the conjugacy class of $\mathbf{x}_1$ is in the same connected component of $\mathcal{T}_G(c)$ as that of (x_0,y_0,z_1), say, and the conjugacy class of $\mathbf{x}_2$ is in the same connected component of $\mathcal{T}_G(c)$ as that of (x_0,y_0,z_2), where z_1 and z_2 lie in L and differ by an element of $Z^0_G(x_0,y_0)$. Hence the classes of (x_0,y_0,z_1) and (x_0,y_0,z_2) are in the same connected component of $\mathcal{T}_G(c)$, as are those $\mathbf{x}_1$ and $\mathbf{x}_2$. Thus ψ is injective on the set of components, and its image is contained in the set of elements in $\pi_0(Z_G(x_0,y_0))$ of order k. To see that it is surjective, choose any element of order k in $\pi_0(Z_G(x_0,y_0))$, represented by $[z_0]$, say. Then (x_0,y_0,z_0) is a c-triple of order k, and by construction $\psi(x_0,y_0,z_0)=[z_0]$. □

COROLLARY 8.3.2. *If k divides at least one of the $g_{\overline{a}}$ for $\overline{a}\in\widetilde{\Delta}_c$, then there are exactly $\varphi(k)$ components of $\mathcal{T}_G(c)$ of order k in G.*

8.4. Proof of Parts 1,2,3 of Theorem 1.5.1 for $\langle C\rangle$ cyclic

We assume that $\langle C\rangle$ is cyclic. Part 1 of Theorem 1.5.1 is contained in Proposition 8.2.2. Part 2 of Theorem 1.5.1 is contained in Proposition 8.2.5. The first statement of Part 3 of Theorem 1.5.1 is Corollary 8.3.2. Let X be a component of $\mathcal{T}_G(c)$ of order k. Let $S=S^{w_c}(\overline{\mathbf{g}},k)$. Since X is the quotient of $S\times S\times S$ by a finite group, Part (1) implies that $d_X = \frac{1}{3}\dim X + 1$ is equal to the number of $\overline{a}$ such that $k|g_{\overline{a}}$. By the first statement of Part (3) and Lemma 3.8.6,

$$\sum_X d_X = \sum_{\overline{a}\in\widetilde{\Delta}_c} g_{\overline{a}} = \sum_{\overline{a}\in\widetilde{\Delta}_c} n_{\overline{a}} g_a = g.$$

8.5. Proof of Part 4 of Theorem 1.5.1 for $\langle C\rangle$ cyclic

THEOREM 8.5.1. *Let X be a component of $\mathcal{T}_G(c)$. Associated to X is the torus S and the group $L = DZ(S)$. Decompose $L = L_0 \times L'$ where L_0 is either trivial or is a simple group not of A-type for any $n \geq 1$. If L_0 is trivial, the map $\overline{p}$ of Theorem 2.3.1 induces a homeomorphism from*

$$(\overline{S} \times \overline{S} \times S)/W(S,G)$$

to X. If L_0 is not trivial, then the map $\overline{p}$ of Theorem 2.3.1 induces a homeomorphism from

$$(\overline{S} \times \overline{S} \times (S/\mathcal{K}))/W(S,G)$$

to X where $\mathcal{K}$ is a subgroup of order at most 2 in $\mathcal{C}L_0$.

PROOF. Let $F = S \cap L$. There is an action of $F \times F \times F$ on the set of all conjugacy classes of rank zero c-triples in L. By Lemma 8.1.10 and Lemma 8.1.11, the action of F on the first two factors is trivial. We analyze the action of F on the last factor. Write $c = c_0 c'$ with $c_0 \in \mathcal{C}L_0$ and $c' \in \mathcal{C}L'$. There is an inclusion $S \cap L \subseteq \mathcal{C}L$. Since L' decomposes as a product of groups of A-type. It is easy to see that $\mathcal{C}L'$ acts freely on the moduli space of c'-triples in L'. According to Lemma 8.1.12 the stabilizer of a conjugacy class of c_0-triples in L_0 is the intersection $\mathcal{C}L_0 \cap S^{w_{c_0}}$, which is a subgroup $\mathcal{K}_0 \subseteq \mathcal{C}L_0$ of index given by the gcd of the quotient coroot integers for L_0. The two possibilities for L_0 when $\dim(S^{w_c}) > 0$ are L_0 of type C_2 or D_6. In both these cases, $\mathcal{K}_0$ is of order 2 and may be described explicitly. We let $\mathcal{K}$ be the intersection $S \cap \mathcal{K}_0 \subseteq S \cap L$. Clearly, $\mathcal{K}$ has order at most 2, and is trivial if L_0 is trivial.

Thus, the stabilizer in F^3 of any c-triple of rank zero in L is equal to $F \times F \times \mathcal{K}$ where $\mathcal{K}$ is trivial if L_0 is trivial, and $\mathcal{K}$ has order at most 2 if L_0 is non-trivial. Finally, it follows from Lemma 8.1.9 that the Weyl group of S acts trivially on the set of conjugacy classes of rank zero c-triples in L. The theorem now follows from Theorem 2.3.1. □

The possibilities for $\mathcal{K}$ and the component X in case L_0 is not trivial are as follows:

- If G is of type B_n, and hence $L_0 = L$ is of type C_2, then $\mathcal{K} = \mathcal{C}L = S \cap L$ has order 2, and $X = (\overline{S} \times \overline{S} \times \overline{S})/W(S,G)$.
- If G is of type C_{2n}, and hence L_0 is of type C_2, then $\mathcal{K}$ is trivial and $X = (\overline{S} \times \overline{S} \times S)/W(S,G)$.
- If G is of type D_{2n}, and hence L_0 is of type D_6, then $\mathcal{K}$ is trivial and $X = (\overline{S} \times \overline{S} \times S)/W(S,G)$.
- If G is of type E_7, and hence $L_0 = L$ is of type D_6, then $\mathcal{K} = S \cap L$ has order 2 and $X = (\overline{S} \times \overline{S} \times \overline{S})/W(S,G)$.

This explicit list completes the proof of Theorem 1.5.1 in case $\langle C\rangle$ is cyclic.

CHAPTER 9

The tori $\overline{S}(k)$ and $\overline{S}^{w_C}(\overline{\mathbf{g}}, k)$ and their Weyl groups

Let Φ be an irreducible but possibly non-reduced root system on a vector space V with set of simple roots Δ. Let $A \subset V$ be the alcove associated with Δ, and let d be the highest root.

DEFINITION 9.0.1. Fix a positive integer n_0 and let $\mathbf{n}\colon \widetilde{\Delta} \to \mathbf{N}$ be the function $n_0\mathbf{g}$. Fix an integer $k > 1$ dividing at least one of the integers $\{\mathbf{n}(a)\}_{a\in\widetilde{\Delta}}$. Let

$$\widetilde{I}(\mathbf{n}, k) = \{a \in \widetilde{\Delta} : k \nmid \mathbf{n}(a)\}.$$

Note that $\widetilde{I}(\mathbf{g}, k) = \widetilde{I}(k)$ as previously defined. In general, if we factor k as dk' where $d = \gcd(k, n_0)$, then $\widetilde{I}(\mathbf{n}, k) = \widetilde{I}(k')$. Let $f(\mathbf{n}, k) \subset A$ be defined by

$$f(\mathbf{n}, k) = \{\, \tilde{x} \in A : a(\tilde{x}) \in \mathbf{Z} \text{ for all } a \in \widetilde{I}(\mathbf{n}, k)\,\}.$$

Let $\hat{f}(\mathbf{n}, k)$ be the affine space generated by $f(\mathbf{n}, k)$ and let $V(\mathbf{n}, k)$ be the linear space parallel to $\hat{f}(\mathbf{n}, k)$. Notice that if $n_0 = 1$ and if $\Phi = \Phi(T, G)$ so that $V = it$, then $f(\mathbf{n}, k) = f(k)$ and $V(\mathbf{n}, k) = it(k)$ as defined in Theorem 1.4.1. More generally, if we factor k as dk' where $d = \gcd(k, n_0)$, then $f(\mathbf{n}, k) = f(k')$ and $V(\mathbf{n}, k) = V(k')$.

9.1. A root system on $\hat{f}(k)$

Let $k \geq 1$ be an integer dividing at least one of the g_a. We denote by $\Phi(k)$ the subroot system of Φ consisting of all roots which annihilate $V(k)$. Suppose that Φ is reduced and let Φ^+ be the set of positive roots for Φ. Recall from Proposition 5.5.5 and Claim 5.5.6 that $\Phi^+(k) = \Phi^+ \cap \Phi(k)$ is a set of positive roots for $\Phi(k)$ and that $\Delta(k)$, the set of simple roots determining this set of positive roots, is given by $\Delta(k) = I(k) \cup \{b\}$ for some root $b \in \Phi(k)$, where $I(k) = \widetilde{I}(k) \cap \Delta$. Furthermore, the root system $\Phi(k)$ is irreducible and d is its highest root. Writing $d^\vee = \sum_{a\in I(k)} g'_a a^\vee + g'_b b^\vee$, we have that $g'_b = k$. Furthermore, for every $a \in I(k)$ the coroot integer g'_a is not divisible by k.

We let U denote the subspace of V spanned by the coroots inverse to the roots in $\Phi(k)$. Then $U = V(k)^\perp$ inside V.

LEMMA 9.1.1. *Let $\tilde{x}_0 = \hat{f}(k) \cap U$ and let $A(k)$ be the alcove in U determined by the set of simple roots $\Delta(k)$ for $\Phi(k)$. Then $\tilde{x}_0$ is the vertex of the alcove $A(k)$ for $\Phi(k)$ opposite the face defined by $\{b = 0\}$.*

PROOF. Each $a \in I(k)$ vanishes on $\hat{f}(k)$ and hence on $\tilde{x}_0$. Similarly, $d(\tilde{x}_0) = 1$. This proves that $\tilde{x}_0$ is the vertex of $A(k)$ opposite the face $\{b = 0\}$. $\square$

LEMMA 9.1.2. *Suppose that $g \in W_{\rm aff}(\Phi)$ and that the differential w of g normalizes $V(k)$. Then there is an element $w' \in W(\Phi(k))$ such that $w'w$ normalizes $\hat{f}(k)$.*

PROOF. Suppose first that Φ is reduced. Since w normalizes $V(k)$, it normalizes U and hence it also normalizes the root system $\Phi(k)$ on U. Thus, $w(\tilde{x}_0)$ is the vertex of an alcove $A'(k)$ for $\Phi(k)$ containing the origin. There is an element $w' \in W(\Phi(k))$ such that $w'(A'(k)) = A(k)$. Then $w'w$ is an automorphism of $A(k)$. Since b is the unique element of $\widetilde{\Delta}(k)$ for which the coroot integer g_b' is divisible by k, the root b is fixed by any automorphism of $\widetilde{\Delta}(k)$. Thus, $\tilde{x}_0$ is fixed by $w'w$. Since $w'w$ fixes $\tilde{x}_0$ and normalizes $V(k)$ it normalizes $\hat{f}(k)$.

If Φ is not reduced, then Φ is of type BC_n for some n. The fundamental relation among the coroots in this case is of the form

$$2\tilde{a}^\vee + 2a_1^\vee + \cdots + 2a_{n-1}^\vee + a_n^\vee$$

where $\tilde{a} = -d$. Thus, $I(2) = \widetilde{I}(2) = \{a_n\}$. (The difference here is that the coefficient of $\tilde{a}^\vee$ is not one, as it is for reduced root systems.) This implies that $f(2)$ is the face of A given by the equation $\{a_n = 0\}$, and so $\hat{f}(2)$ is a linear subspace. Hence $\hat{f}(2) = V(2)$, and the statement of the lemma is obvious. □

COROLLARY 9.1.3. *Every element $\overline{w} \in W(V(k), \Phi)$ has a representative $w \in N_{W(\Phi)}(\hat{f}(k))$.*

LEMMA 9.1.4. *Suppose that Φ is reduced. Any $\tilde{x}' \in \hat{f}(k)$ is equivalent under the action of $W_{\rm aff}(\Phi)$ to a point in $f(k)$.*

PROOF. We may assume that $\Phi = \Phi(T, G)$ for some simple group G with maximal torus T. We set $S(k) = \exp(\mathfrak{t}(k))$ and let $L(k) = DZ(S(k))$. Let $x_0 = \exp(2\pi i\tilde{x}_0) \in L(k)$, where (x_0, y_0, z_0) is a commuting triple of order k and rank zero in $L(k)$. Then $Z_G(x_0, y_0, z_0)$ has $S(k)$ as a maximal torus. Let $\tilde{x}' \in \hat{f}(k)$ and let $x' = \exp(2\pi i\tilde{x}')$. Then (x', y_0, z_0) is a commuting triple of order k in G. According to Part 3 of Proposition 5.5.1, the point x' is conjugate in G and hence in $W(\Phi)$ to a point which is the exponential of a point in $2\pi i f(k)$. Hence, $\tilde{x}'$ is conjugate under $W_{\rm aff}(\Phi)$ to a point of $f(k)$. □

COROLLARY 9.1.5. *Suppose that Φ is irreducible but not necessarily reduced. The alcove decomposition of V induces a decomposition of $\hat{f}(k)$ into compact convex regions with disjoint interiors. Let $\mathcal{A}(k)$ be this decomposition. Then $f(k) \in \mathcal{A}(k)$. Moreover, the action of $N_{W_{\rm aff}(\Phi)}(\hat{f}(k))$ on $\hat{f}(k)$ preserves the decomposition $\mathcal{A}(k)$ and is transitive on $\mathcal{A}(k)$.*

PROOF. First consider the case when Φ is reduced. Clearly, $N_{W_{\rm aff}(\Phi)}(\hat{f}(k))$ acts on $\hat{f}(k)$ normalizing $\mathcal{A}(k)$ and $f(k)$ is one of the elements of this decomposition. Suppose that A_1 is an element of this decomposition. Let $\tilde{x}'$ be an interior point of A_1. According to the previous lemma, there is an element $g \in W_{\rm aff}(\Phi)$ which conjugates $\tilde{x}'$ to a point of $f(k)$. Since there are only finitely many elements of $W_{\rm aff}(\Phi)$ which conjugate A_1 so as to meet $f(k)$, it follows that there is some element $g \in W_{\rm aff}(\Phi)$ which conjugates an open subset of A_1 into $f(k)$. This element normalizes $\hat{f}(k)$ and hence sends A_1 onto $f(k)$. This proves the transitivity statement in the reduced case.

If Φ is not reduced, then it is of type BC_n for some n, $k = 2$ and $\hat{f}(2)$ is the linear subspace spanned by the coroots of the subsystem BC_{n-1}. The induced alcove decomposition of $\hat{f}(2)$ is exactly the alcove decomposition for BC_{n-1}. Since $N_{W_{\text{aff}}(\Phi)}(\hat{f}(2))$ contains the affine Weyl group of BC_{n-1}, the lemma is clear in this case. □

LEMMA 9.1.6. *If $g \in W_{\text{aff}}(\Phi)$ has the property that there is an element of the decomposition $\mathcal{A}(k)$ normalized by g, then $g|\hat{f}(k)$ is the identity.*

PROOF. Let A_1 be an element of $\mathcal{A}(k)$ such that $g(A_1) = A_1$. Let A be an alcove for Φ containing A_1 in its closure, and let $B = g(A)$. Then B also contains A_1, and hence $A \cap B$ contains A_1. But the unique element of $W_{\text{aff}}(\Phi)$ taking A to B fixes pointwise the intersection $A \cap B$. Thus, $g|A_1$ is the identity, and consequently $g|\hat{f}(k)$ is the identity. □

COROLLARY 9.1.7. *Define $W_{\text{aff}}(\hat{f}(k))$ be $N_{W_{\text{aff}}(\Phi)}(\hat{f}(k))/Z_{W_{\text{aff}}(\Phi)}(\hat{f}(k))$. Then $W_{\text{aff}}(\hat{f}(k))$ acts on $\hat{f}(k)$ as a group of affine isometries. It acts simply transitively on the elements of the decomposition $\mathcal{A}(k)$.*

PROPOSITION 9.1.8. *The reflections of $\hat{f}(k)$ in the walls of $f(k)$ are realized by elements in $W_{\text{aff}}(\hat{f}(k))$. These reflections generate $W_{\text{aff}}(\hat{f}(k))$ which is thus a Coxeter group with fundamental domain $f(k)$.*

PROOF. Consider an element A_1 for the decomposition $\mathcal{A}(k)$ which shares a codimension-one wall with $f(k)$. Let g be an element of $N_{W_{\text{aff}}(\Phi)}(\hat{f}(k))$ which sends $f(k)$ to A_1. This affine isometry of $\hat{f}(k)$ fixes the intersection of $f(k)$ and A_1, which is a codimension-one affine subspace. Thus, it is a reflection in this subspace. This shows that the reflections in the walls of $f(k)$ are elements of $W_{\text{aff}}(\hat{f}(k))$. Since $W_{\text{aff}}(\hat{f}(k))$ acts simply transitively on the elements in the decomposition $\mathcal{A}(k)$, it follows that these reflections generate $W_{\text{aff}}(\hat{f}(k))$. □

COROLLARY 9.1.9. *There is a reduced root system $\Phi(V(k))$ on $V(k)$ and a vertex v of $f(k)$ such that, using v to identify $\hat{f}(k)$ with $V(k)$, the affine Weyl group of $\Phi(V(k))$ is identified with $W_{\text{aff}}(\hat{f}(k))$.*

PROOF. The follows immediately from the fact, established in the previous proposition, that $W_{\text{aff}}(\Phi(\hat{f}(k))$ is a Coxeter group. □

COROLLARY 9.1.10. *We define an embedding of $\widetilde{\Delta}^\vee - \widetilde{I}^\vee(k)$ into $V(k)$ by sending $a^\vee$ to $\pi_k(a^\vee)$, where π_k is the orthogonal projection onto $V(k)$. Up to positive multiples $\pi_k(\widetilde{\Delta}^\vee - I^\vee(k)) \subset V(k)$ is the set of extended coroots for the root system $\Phi(V(k))$. In particular, $W(V(k), \Phi)$ is the group generated by reflections in the the walls orthogonal to $\pi_k(a^\vee)$ for $a \in \widetilde{\Delta}^\vee - \widetilde{I}^\vee(k)$.*

PROOF. Since the walls of the alcove $f(k) \subset \hat{f}(k)$ are the subspaces of $\hat{f}(k)$ orthogonal to the $\pi_k(a^\vee)$ for $a \in \widetilde{\Delta} - \widetilde{I}(k)$, the first statement is clear. Clearly, the image under the differential of $W_{\text{aff}}(\hat{f}(k)) = N_{W_{\text{aff}}(\Phi)}(\hat{f}(k))/Z_{W_{\text{aff}}(\Phi)}(\hat{f}(k))$ is contained in $W(V(k), \Phi)$. By Corollary 9.1.3 this map is onto. By Proposition 9.1.8, $W_{\text{aff}}(\hat{f}(k))$ is generated by the reflections in the walls of $f(k)$. The image under the differential of these reflections is the set of reflections of $V(k)$ in the $\pi_k(a^\vee)$ for $a \in \widetilde{\Delta} - \widetilde{I}(k)$. □

LEMMA 9.1.11. *The lattice generated by $\pi_k(a^\vee)$ for $a \in \widetilde{\Delta} - \widetilde{I}(k)$ is the image under the orthogonal projection of the coroot lattice $Q^\vee(\Phi)$.*

PROOF. Since $\widetilde{\Delta}^\vee$ spans $Q^\vee(\Phi)$ and since $\pi_k(a^\vee) = 0$ if $a \in \widetilde{I}(k)$, this is clear. □

PROPOSITION 9.1.12. *The coroot lattice of $\Phi(V(k))$ is $\pi_k(Q^\vee(\Phi))$.*

PROOF. The coroot lattice of $\Phi(V(k))$ is identified with the group of translations of $\hat{f}(k)$ which occur as restrictions of elements of $W_{\text{aff}}(\Phi)$ normalizing $\hat{f}(k)$.

Let $\lambda \in Q^\vee(\Phi)$. Translation by λ carries $\hat{f}(k)$ to an affine subspace $\hat{f}_1$ of V which meets U in a point of the form $\tilde{x}_0 + \zeta$ where ζ is an element in the dual to the root lattice of $\Phi(k)$. In particular, $[\zeta] \in U/Q^\vee(\Phi(k))$ is an element of $\mathcal{C}\Phi(k) = P^\vee(\Phi(k))/Q^\vee(\Phi(k))$. But since $\tilde{x}_0$ is the vertex of $A(k)$ opposite the wall $\{b = 0\}$, and b is the unique element in $\widetilde{\Delta}(\Phi(k))$ whose coroot integer is divisible by k, it follows that the action of $\mathcal{C}\Phi(k)$ on $A(k)$ fixes $\tilde{x}_0$. Thus there is an element $w \in W_{\text{aff}}(\Phi(k))$ such that $w(\tilde{x}_0 + \zeta) = \tilde{x}_0$. The composition of translation by λ followed by w is an element of $W_{\text{aff}}(\Phi)$ normalizing $\hat{f}(k)$ and acting on it by translation by $\pi_k(w\lambda) = \pi_k(\lambda)$. This shows that $\pi_k(Q^\vee(\Phi))$ is contained in the coroot lattice of $\Phi(V(k))$.

Conversely, suppose that $w \in W_{\text{aff}}(\Phi)$ normalizes $\hat{f}(k)$ and acts on it by a pure translation. We write $w(x) = w_0(x) + \lambda$ where $\lambda \in Q^\vee(\Phi)$ and w_0 is in the Weyl group of Φ. The element w_0 normalizes $V(k)$ and hence U so that $w_0(\tilde{x}_0) = \tilde{x}_0 - \zeta$ for some $\zeta \in U$. Thus, $w(\tilde{x}_0) = \tilde{x}_0 + (\lambda - \zeta)$. Since this element restricts to $\hat{f}(k)$ to give a translation, its restriction is translation by $\pi_k(\lambda - \zeta)$. Since $\zeta \in U = \text{Ker}(\pi_k)$, we have $\pi_k(\lambda - \zeta) = \pi_k(\lambda) \in \pi_k(Q^\vee(\Phi))$. □

THEOREM 9.1.13. *The reduced root system $\Phi(V(k))$ on $V(k)$ has Weyl group $W(V(k), \Phi)$. It has coroot lattice equal to $\pi_k(Q^\vee(\Phi))$. If $\dim V(k) \geq 1$, then $\{\pi_k(a^\vee)\}_{a \in \tilde{\Delta} - \tilde{I}(k)}$ as an extended set of coroots for $\Phi(V(k))$.*

PROOF. Everything except the last statement is contained in Corollary 9.1.10 and Proposition 9.1.12. Since $\pi_k(a^\vee) \in \pi_k(Q^\vee(\Phi))$ and since these form, up to positive multiples, a set of extended coroots, to complete the proof we need only see that the $\pi_k(a^\vee)$ are indivisible elements of $\pi_k(Q^\vee(\Phi))$. This will follow from the next lemma.

LEMMA 9.1.14. *Writing $b^\vee = \sum_{a \in \Delta - I(k)} m_a a^\vee + \sum_{a \in I(k)} n_a a^\vee$, one of the following holds:*

(1) *The cardinality of $\Delta - I(k)$ is one and the unique m_a is one;*
(2) *There are at least two $a \in \Delta - I(k)$ for which $m_a = 1$.*

PROOF. We have

$$d = \sum_{a \in I(k)} g'_a a^\vee + k \left(\sum_{a \in \Delta - I(k)} m_a a^\vee + \sum_{a \in I(k)} n_a a^\vee \right) = \sum_{a \in \Delta} g_a a^\vee.$$

Thus, for $a \in \Delta - I(k)$, $g_a = k m_a$, and the remaining g_a are not divisible by k. The result then follows from Lemma 3.8.5. □

Now let us return to the proof of the theorem. Since $I(k)\cup\{b\}$ is a set of simple roots for $\Phi(k)$, we have an exact sequence

$$0\to\sum_{a\in I(k)}\mathbf{Z}(a^\vee)\oplus\mathbf{Z}(b^\vee)\to\sum_{a\in\Delta}\mathbf{Z}(a^\vee)\to\pi_k(Q^\vee)\to 0.$$

We can rewrite this sequence as

$$0\to\mathbf{Z}(b^\vee)\to\sum_{a\in\Delta-I(k)}\mathbf{Z}(a^\vee)\to\pi_k(Q^\vee(\Phi))\to 0.$$

According to Lemma 9.1.14, the image of $b^\vee$ in $\pi_k(Q^\vee(\Phi))$ is $\sum_{a\in\Delta-I(k)}m_a a^\vee$ where either the cardinality of $\Delta-I(k)$ is one or there are at least two $a\in\Delta-I(k)$ for which $m_a=1$. In the first case $V(k)$ is a point. In the second case, it is easy to see that the image of each $a^\vee\in\Delta-I(k)$ is indivisible in $\pi_k(Q^\vee(\Phi))$. $\square$

Let us return to the case of a general function $\mathbf{n}=n_0\mathbf{g}$. Then $\widetilde{I}(\mathbf{n},k)=\widetilde{I}(k')$ and $V(\mathbf{n},k)=V(k')$ where $k'=k/\gcd(n_0,k)$, and thus $\Phi(V(\mathbf{n},k))=\Phi(V(k'))$. The following is then an immediate corollary of Theorem 9.1.13.

COROLLARY 9.1.15. *The Weyl group of $V(\mathbf{n},k)$ in Φ is the Weyl group of the reduced root system $\Phi(V(\mathbf{n},k))$. The coroot lattice of $\Phi(V(\mathbf{n},k))$ is $\pi_k(Q^\vee(\Phi))$. If* $\dim V(\mathbf{n},k)\geq 1$, *then $\{\pi_k(a^\vee)\}_{a\in\widetilde{\Delta}-\widetilde{I}(\mathbf{n},k)}$ is an extended set of simple coroots of* $\Phi(V(\mathbf{n},k))$.

9.2. Completion of the proof of Theorem 1.4.1

To complete the proof of Theorem 1.4.1 we must show that the torus $\overline{S}(k)$ and the Weyl group $W(S(k),G)$ are as described in Part 5 of the statement of that theorem. This is immediate by applying Theorem 9.1.13 to the root system $\Phi=\Phi(T,G)$ using the fact that setting $V=i\mathfrak{t}$ we have $V(k)=i\mathfrak{t}(k)$ and $2\pi i\pi_k(Q^\vee)\subset\mathfrak{t}(k)$ is the fundamental group of $\overline{S}(k)$

9.3. Completion of the proof of Theorem 1.5.1 in case $\langle C\rangle$ is cyclic

Let us assume that $\langle C\rangle$ is cyclic and generated by c. It remains to establish that $\overline{S}^{w_c}(\overline{\mathbf{g}},k)$ and $W(S^{w_c}(\overline{\mathbf{g}},k),G)$ are as given in Part 5 of the statement of that theorem.

LEMMA 9.3.1. *Every element in $W(S^{w_c}(\overline{\mathbf{g}},k),G)$ has a representative in the Weyl group of G which normalizes S^{w_c}.*

PROOF. Set $S=S^{w_c}(\overline{\mathbf{g}},k)$ and set $L=DZ(S)$. According to Theorem 2.3.1, there is a c-triple (x_0,y_0,z_0) of rank zero in L, and S is a maximal torus of $Z_G(x_0,y_0,z_0)$. By Corollary 6.1.8 we can assume that (x_0,y_0) is a c-pair in L in weak normal form with respect to $T\cap L$.

Let $g\in N_G(T)$ normalize S. Then g also normalizes L and $g(x_0,y_0,z_0)g^{-1}$ is another c-triple of rank zero and order k in L. Thus, by Proposition 8.1.7 this triple is conjugate by an element $h\in L$ to a triple of the form (x_0,y_0,z_0^ℓ) for some $\ell\in\mathbf{Z}$. In particular, $hg(x_0,y_0)g^{-1}h^{-1}=(x_0,y_0)$. Since (x_0,y_0,z_0) is a triple of rank zero in L, the centralizer $Z_L^0(x_0,y_0)$ is a torus. Suppose that a root of G with respect to T is one on $S\cdot x_0$ and $S\cdot y_0$. This root then is one on S, and hence is a root of L with respect to $T\cap L$ which is one on x_0 and y_0. But we have just seen that there are no such roots. Thus, $Z_G^0(S\cdot x_0,S\cdot y_0)$ is also a torus. Since (x_0,y_0) is a c-pair in weak normal form, Lemma 7.2.1 implies that $S^{w_c} \subseteq Z_G^0(x_0, y_0)$ is a maximal torus. Because hg fixes (x_0, y_0) and normalizes S, it normalizes $Z_G^0(S \cdot x_0, S \cdot y_0) = S^{w_c}$. Clearly, the image in $W(S,G)$ of hg is equal to that of g. $\square$

Let us consider the root system $\Phi^{\rm proj}(w_c)$ on $i\mathfrak{t}^{w_c}$. According to Proposition 6.2.6 it is irreducible, but possibly not reduced. Let $\overline{\mathbf{g}}\colon \widetilde{\Delta}_c \to \mathbf{N}$ be the function defined by $\overline{\mathbf{g}}(\overline{a}) = g_{\overline{a}}$. Applying Definition 9.0.1 to $\mathfrak{t}^{w_c}$ and the function $\overline{\mathbf{g}}$ produces the subspace $i\mathfrak{t}^{w_c}(\overline{\mathbf{g}}, k)$ as given in Definition 8.2.1.

We know by Proposition 6.6.4 that the elements $\widetilde{\Delta}_c^\vee$ are the extended set of coroots for the root system $\Phi(w_c)$ of Definition 6.6.5. Of course, $\Phi(w_c)$ has the same Weyl group and coroot lattice as $\Phi^{\rm proj}(w_c)$. In particular, its coroot lattice is the orthogonal projection $Q^\vee$ into $i\mathfrak{t}^{w_c}$. Corollary 9.1.15 applied to $\Phi(w_c)$ and $\widetilde{\Delta}_c$ then implies that the lattice generated by the images under orthogonal projection to $i\mathfrak{t}^{w_c}(\overline{\mathbf{g}}, k)$ of $\widetilde{\Delta}_c^\vee - \widetilde{I}_c^\vee(k)$ is exactly the image under orthogonal projection of $Q^\vee$, and that the Weyl group of $i\mathfrak{t}^{w_c}(\overline{\mathbf{g}}, k)$ with respect to the root system $\Phi^{\rm proj}(w_c)$, or equivalently with respect to the root system $\Phi(w_c)$, is the group generated by reflections in the walls orthogonal to the images under orthogonal projection of the $a^\vee$ for $a \in \widetilde{\Delta}_c - \widetilde{I}_c(k)$. By Proposition 6.5.5, the Weyl group of $\Phi^{\rm proj}(w_c)$ is equal to $W(i\mathfrak{t}^{w_c}, \Phi(T,G))$, which is identified $W(\mathfrak{t}^{w_c}, G)$. Thus, we see that the group generated by the reflections in the walls orthogonal to the images under orthogonal projection of the $a^\vee$ for $a \in \widetilde{\Delta}_c - \widetilde{I}_c(k)$ is equal to the subgroup of $W(i\mathfrak{t}^{w_c}(\overline{\mathbf{g}}, k), \Phi)$ realized by elements normalizing both $\mathfrak{t}^{w_c}(\overline{\mathbf{g}}, k)$ and $\mathfrak{t}^{w_c}$. By the previous lemma, this is the entire Weyl group $W(i\mathfrak{t}^{w_c}(\overline{\mathbf{g}}, k), \Phi) = W(\mathfrak{t}^{w_c}, G)$.

9.4. The generalized Cartan matrix associated to $\widetilde{\Delta}^\vee - \widetilde{I}^\vee(\mathbf{n}, k) \subset \mathfrak{t}(\mathbf{n}, k)$

Let Φ be an irreducible, but possibly non-reduced, root system on V with extended set of simple coroots $\widetilde{\Delta}$. Let $\widetilde{D}$ be the extended coroot diagram of Φ. Fix a function $\mathbf{n}\colon \widetilde{\Delta} \to \mathbf{N}$ of the form $n_0\mathbf{g}$ for some positive integer n_0 and fix $k \geq 1$ dividing at least one of the integers $n_0 g_a$. Let $\ell\colon \widetilde{\Delta}^\vee \to \mathbf{R}^+$ be the length function determined by the inner product on $\mathfrak{t}$. Fix $k \geq 1$ dividing one of the integers $n_0 g_a$. Let π_k denote orthogonal projection from V to $V(\mathbf{n}, k)$. Consider the image under orthogonal projection of $\widetilde{\Delta}^\vee - \widetilde{I}^\vee(\mathbf{n}, k)$. According to Theorem 9.1.13, π_k embeds $\widetilde{\Delta}^\vee - \widetilde{I}^\vee(\mathbf{n}, k) \subset V(\mathbf{n}, k)$ as a set of extended coroots for a reduced root system $\Phi(V(\mathbf{n}, k))$. In Definition 1.7.2 we defined a diagram $\widetilde{D}(\mathbf{n}, k)$ with nodes $\widetilde{\Delta}^\vee - \widetilde{I}^\vee(\mathbf{n}, k)$. On the other hand, by Corollary 9.1.15, the set $\{\pi_k(a^\vee) : a^\vee \in \widetilde{\Delta}^\vee - \widetilde{I}^\vee(\mathbf{n}, k)\} \subset \Phi^\vee(V(\mathbf{n}, k))$ is an extended set of simple coroots for the root system $\Phi(V(\mathbf{n}, k))$, whose Cartan integers are given by

$$n(\pi_k(a^\vee), \pi_k(b^\vee)) = 2\frac{\langle \pi_k(a^\vee), \pi_k(b^\vee)\rangle}{\langle \pi_k(b^\vee), \pi_k(b^\vee)\rangle}.$$

Let $\widetilde{D}_0(\mathbf{n}, k)$ be the corresponding extended coroot diagram. Orthogonal projection identifies the nodes of $\widetilde{D}(\mathbf{n}, k)$ with those of $\widetilde{D}_0(\mathbf{n}, k)$.

THEOREM 9.4.1. *Under the above identification of the nodes, the diagrams $\widetilde{D}(\mathbf{n}, k)$ and $\widetilde{D}_0(\mathbf{n}, k)$ coincide.*

Proposition 1.7.3 is an immediate corollary of Theorem 9.4.1. We will prove Proposition 1.7.1 in the course of proving Theorem 9.4.1.

Since $\widetilde{I}(\mathbf{n},k) = \widetilde{I}(k')$ where $k' = k/\gcd(k, n_0)$, without loss of generality, for the rest of this section, we assume that $\mathbf{n} = \mathbf{g}$ and drop it from the notation. In particular, we denote $\widetilde{D}(\mathbf{g},k)$ and $\widetilde{D}_0(\mathbf{g},k)$ by $\widetilde{D}(k)$ and $\widetilde{D}_0(k)$.

If $k = 1$, then $\widetilde{I}(k) = \emptyset$, π_k is the identity and there is nothing to prove. Thus, from now on we assume that $k > 1$. This implies that Φ is not of type A and hence $\widetilde{D}$ is contractible. Let $\widetilde{D}'(k)$ be the sub-diagram of $\widetilde{D}(\Phi) = \widetilde{D}$ spanned by the nodes of $\widetilde{I}(k)$.

LEMMA 9.4.2. *Let $a^\vee, b^\vee \in \widetilde{\Delta}^\vee - \widetilde{I}^\vee(k)$.*

(1) *The node of $\widetilde{D}$ corresponding to $a^\vee$ is not connected in $\widetilde{D}$ to any node of $\widetilde{D}'(k)$ if and only if $a^\vee \in \mathfrak{t}(k)$.*
(2) *If $a^\vee \in \mathfrak{t}(k)$, then $\langle a^\vee, b^\vee\rangle = \langle \pi_k(a^\vee), \pi_k(b^\vee)\rangle$.*
(3) *Suppose that $a^\vee, b^\vee$ are adjacent nodes of $\widetilde{D}$. Then $\langle \pi_k(a^\vee), \pi_k(b^\vee)\rangle = \langle a^\vee, b^\vee\rangle$.*
(4) *If $a^\vee$ and $b^\vee$ are not adjacent nodes of $\widetilde{D}$ and if they are not connected to a common component of $\widetilde{D}'(k)$, then $\langle \pi_k(a^\vee), \pi_k(b^\vee)\rangle = 0$.*
(5) *If $a^\vee$ and $b^\vee$ are not adjacent in $\widetilde{D}$, but $a^\vee$ and $b^\vee$ are connected to a common component of $\widetilde{D}'(k)$, then $\langle \pi_k(a^\vee), \pi_k(b^\vee)\rangle < 0$.*

PROOF. Let $U \subset V$ be the subspace $V(k)^\perp$. It is the subspace spanned of the coroots inverse to the roots represented by nodes of $\widetilde{D}'(k)$. As such, it decomposes as an orthogonal sum of the subspaces U_i indexed by the connected components of $\widetilde{D}'(k)$. The factor U_i corresponding to a component is the subspace of $\mathfrak{u}$ spanned by the coroots inverse to the roots represented by the nodes of that component.

The coroot $a^\vee$ is contained in $V(k)$ if and only if it is orthogonal to all the coroots inverse to the roots corresponding to nodes of $\widetilde{D}'(k)$. This is equivalent to the node of $\widetilde{D}$ corresponding to $a^\vee$ not being connected in $\widetilde{D}$ to any node of $\widetilde{D}'(k)$.

If $a^\vee \in V(k)$, then $\pi_k(a^\vee) = a^\vee$ and $\langle \pi_k(a^\vee), \pi_k(b^\vee)\rangle = \langle a^\vee, b^\vee\rangle$. Thus the second item is clear.

If $a^\vee$ and $b^\vee$ are adjacent nodes of $\widetilde{D}$, then since $\widetilde{D}$ is contractible, $a^\vee$ and $b^\vee$ are not connected to a common component of $\widetilde{D}'(k)$. This means that the projections $\pi_U(a^\vee)$ and $\pi_U(b^\vee)$ are orthogonal, where $\pi_\mathfrak{u}$ denotes orthogonal projection to U. The third item follows.

Suppose that $a^\vee$ and $b^\vee$ are not adjacent and are not connected in $\widetilde{D}$ to a common component of $\widetilde{D}'(k)$. Then there is no factor U_i of U with the property that the orthogonal projections of both $a^\vee$ and $b^\vee$ into U_i are both non-trivial. Thus, $\langle \pi_k(a^\vee), \pi_k(b^\vee)\rangle = \langle a^\vee, b^\vee\rangle = 0$.

Lastly, if $a^\vee$ and $b^\vee$ are not adjacent in $\widetilde{D}$ then $\langle a^\vee, b^\vee\rangle = 0$ and hence $\langle \pi_k(a^\vee), \pi_k(b^\vee)\rangle = -\langle \pi_U(a^\vee), \pi_U(b^\vee)\rangle$. Thus, we complete the proof by showing that, under the hypothesis of Part 5, the inner product $\langle \pi_U(a^\vee), \pi_U(b^\vee)\rangle > 0$. Let U_i be the subspace spanned by the coroots inverse to the roots corresponding to the nodes of the component of $\widetilde{D}'(k)$ connected to both $a^\vee$ and $b^\vee$. By Theorem 3.6.1, the component of $\widetilde{D}'(k)$ connected to both $a^\vee$ and $b^\vee$ is of type A_n for some $n \geq 1$. Furthermore, if $\{a_1^\vee, \ldots, a_n^\vee\}$ is the set of simple coroots for this component given by the nodes of $\widetilde{D}'(k)$, then $a^\vee$ is connected to a unique $a_i^\vee$ and $b^\vee$ is connected to a unique $a_j^\vee$. Thus, $\pi_{U_i}(a^\vee)$ is a negative multiple of the fundamental coweight $\varpi_{a_i}^\vee$ for the root system corresponding to $\widetilde{D}'(k)$, and likewise $\pi_{U_i}(b^\vee)$ is a negative multiple of $\varpi^\vee_{a_j}$. The following computation in A_n shows then that these vectors have positive inner product.

LEMMA 9.4.3. *Let $a_1,\dots,a_n$ be the simple roots in A_n. Order the a_i so that $\langle a_i,a_{i+1}\rangle=-1$ for $1\le i\le n-1$, where $\langle\cdot,\cdot\rangle$ is the standard Weyl invariant inner product. Let ϖ_{a_i} be the fundamental weight corresponding to a_i. Then for $i\le j$,*

$$\langle\varpi_{a_i},\varpi_{a_j}\rangle=\frac{i(n+1-j)}{n+1}.$$

PROOF. This is a straightforward computation. □

COROLLARY 9.4.4. *Let $a^\vee,b^\vee\in\widetilde{\Delta}^\vee-\widetilde{I}^\vee(k)$. Then $\langle\pi_k(a^\vee),\pi_k(b^\vee)\rangle<0$ if and only if either $a^\vee$ and $b^\vee$ correspond to nodes of $\widetilde{D}$ which are adjacent in $\widetilde{D}$, or correspond to nodes of $\widetilde{D}$ which are connected in $\widetilde{D}$ to a common component of $\widetilde{D}'(k)$.*

In particular, the corollary tells us which nodes in $\widetilde{D}_0(k)$ are connected by a bond in the diagram. This agrees with the recipe given in Part 1 of Definition 1.7.2 for $\widetilde{D}(k)$. To determine the multiplicities and directions of the bonds in $\widetilde{D}_0(k)$, we compute the lengths of the $\pi_k(a^\vee)$. Suppose that $a^\vee\in\widetilde{\Delta}^\vee-\widetilde{I}^\vee(k)$. Let $\{a_1^\vee,\dots,a_r^\vee\}\subset\widetilde{I}^\vee(k)$ be the nodes of $\widetilde{D}'(k)$ that a meets. Since by hypothesis $\widetilde{D}$ is contractible, the $a_i^\vee$ are simple coroots of distinct irreducible factors of $\Phi(k)$, and hence are mutually orthogonal. We have $\sum_b g_b b^\vee=0$ and $g_b\equiv 0 \pmod k$ for all $b^\vee\notin\widetilde{I}^\vee(k)$. Since $n(a^\vee,b^\vee)=0$ for all $b^\vee\in\widetilde{I}^\vee(k)-\{a_1^\vee,\dots,a_r^\vee\}$, we have

$$\sum_{i=1}^{r} n(a_i^\vee,a^\vee)g_{a_i}\equiv 0 \pmod k. \tag{9.1}$$

Of course, in the above congruence, each g_{a_i} is not divisible by k.

If $\widetilde{\Delta}^\vee-\widetilde{I}^\vee(k)$ is a single node, then clearly this node is of Type ∞ in the terminology of Proposition 1.7.1. Thus, we can assume that the cardinality of $\widetilde{\Delta}^\vee-\widetilde{I}^\vee(k)$ is at least 2, and hence that $\pi_k(a^\vee)\neq 0$ for every $a^\vee\in\widetilde{\Delta}^\vee-\widetilde{I}^\vee(k)$. This rules out the root systems G_2 and BC_1. Thus, from now on we assume that $\widetilde{D}$ has only single and double bonds.

We shall now complete the proof of Proposition 1.7.1 and the proof that $\widetilde{D}_0(k)=\widetilde{D}(k)$ by examining the various possibilities for k. The projection of $a^\vee\in\widetilde{\Delta}^\vee-\widetilde{I}^\vee(k)$ is a sum of coweights of the form $\varpi^\vee_{a_i}$, where a_i is a simple root in a root system of type A, and we shall tacitly use Lemma 9.4.3 in the calculations below in the case $i=j$.

The case $k=2$.

By Theorem 3.6.1, or direct inspection in the case of BC_n, all the components of $\widetilde{D}'(2)$ are of A_1-type. Number the nodes $a_1^\vee,\dots,a_r^\vee$ of $\widetilde{D}'(2)$ connected to $a^\vee$ in such a way that the nodes $a_1^\vee,\dots,a_s^\vee$ are of different length from $a^\vee$ and $a_{s+1}^\vee,\dots,a_r^\vee$ are of same length as $a^\vee$. We set $t=r-s$. The $a_i^\vee$ are mutually orthogonal. Thus, the length squared of the orthogonal projection of $a^\vee$ into $\mathfrak{u}$ is

$$\left(\frac{t}{4}+\frac{s}{2}\right)|a^\vee|^2.$$

Since the projection of $a^\vee$ into $\mathfrak{u}$ has length less than that of $a^\vee$, it follows that

$$t+2s<4. \tag{9.2}$$

If $s = 0$, then $\ell(a^\vee) = \ell(a_i^\vee)$ for all i, and hence $n(a_i^\vee, a^\vee) = -1$ for all i. From Equation 9.1 we see that t is even, and, by Inequality 9.2, that t is either 0 or 2. If $s = t = 0$, then $a^\vee$ is of Type 1 in the terminology of Proposition 1.7.1 and $a^\vee = \pi_k(a^\vee)$ so that $|\pi_k(a^\vee)| = |a^\vee| = \ell_2(a^\vee)$. If $s = 0$ and $t = 2$, then $a^\vee$ is of Type 2(i) and $|\pi_k(a^\vee|^2 = |a^\vee|^2/2 = \ell_2^2(a^\vee)$.

If $s = 1$ and $a^\vee$ is long, then $n(a_i^\vee, a^\vee) = -1$ for all i and by Equation 9.1 we see that $s+t$ is even. Thus by Inequality 9.2, $s = t = 1$. Thus, $a^\vee$ is of Type 4(i) In this case the length squared of the orthogonal projection of $a^\vee$ into $\mathfrak{u}$ is $(3/4)|a^\vee|^2$ and thus $|\pi_k(a^\vee)|^2 = |a^\vee|^2/4 = \ell_4^2(a^\vee)$.

If $s = 1$ and $a^\vee$ is short then $n(a_1^\vee, a^\vee) = -2$ and $n(a_i^\vee, a^\vee) = -1$ for $i > 1$ and hence by Equation 9.1 it follows that t is even. Inequality 9.2 implies that $t = 0$. Thus, $a^\vee$ is of Type 2(ii) and the length squared of the projection of $a^\vee$ into $\mathfrak{u}$ is $(1/2)|a^\vee|^2$ and thus $\pi_k(a^\vee)|^2 = |a^\vee|^2/2 = \ell_2^2(a^\vee)$.

This shows that all $a^\vee \in \widetilde{\Delta}^\vee - \widetilde{I}^\vee(2)$ are of Type 1, Type 2, Type 4 or Type ∞, and that $\widetilde{D}_0(2) = \widetilde{D}(2)$.

The case $k = 3$.

In this case, again by Theorem 3.6.1, all the components of $\widetilde{D}'(3)$ are of A_2-type. Suppose that $a^\vee \in \widetilde{\Delta}^\vee - \widetilde{I}^\vee(3)$ and that it is connected to nodes $\{a_1^\vee, \ldots, a_r^\vee\}$ of $\widetilde{D}'(3)$. Number the nodes $a_1^\vee, \ldots, a_r^\vee$ of $\widetilde{D}'(3)$ connected to $a^\vee$ in such a way that the nodes $a_1^\vee, \ldots, a_s^\vee$ are of different length from $a^\vee$ and $a_{s+1}^\vee, \ldots, a_r^\vee$ are of same length as $a^\vee$. We set $t = r - s$. Since the $a_i^\vee$ are roots of distinct irreducible components of $\widetilde{D}'(3)$ and hence mutually orthogonal. The length squared of the orthogonal projection of $a^\vee$ into $\mathfrak{u}$ is

$$\left(\frac{2s+t}{3}\right)|a^\vee|^2.$$

Since this length squared is less than $|a^\vee|^2$, the possibilities are $s = 1, t = 0$ or $s = 0, t \leq 2$. By Equation 9.1, we see that $s = 0$ and that t is either 0 or 2. When $s = t = 0$ $a^\vee$ is of Type 1 and $\pi_k(a^\vee) = a^\vee$ so that $\ell_3(a^\vee) = |\pi_k(a^\vee)|$. If $s = 0$ and $t = 2$, then $a^\vee$ is of Type 3 and $|\pi_k(a^\vee)|^2 = |a^\vee|^2/3 = \ell_3^2(a^\vee)$.

This shows that all $a^\vee \in \widetilde{\Delta}^\vee - \widetilde{I}^\vee(3)$ are of Type 1, Type 3, or Type ∞, and that $\widetilde{D}_0(3) = \widetilde{D}(3)$.

The case $k = 4$. Of course, the same kind of general arguments as above can be made, using the fact that the components of $\widetilde{D}'(4)$ are all of type A_1 or A_3. However, the only case where there is a node not of Type ∞ is when Φ is of type E_8. It can be checked directly in this case that all nodes are of Type 4(ii) or (iii) and that the lengths are as stated.

The case $k > 4$. In this case only nodes of Type ∞ arise.

This completes the proof that $\widetilde{D}_0(k) = \widetilde{D}(k)$, and hence of Theorem 9.4.1. In the course of the proof we showed that every node of $\widetilde{\Delta}^\vee - \widetilde{I}^\vee(k)$ is of one of the types listed in Proposition 1.7.1, thus proving that result.

9.5. Proof of Theorem 1.7.4

According to Theorem 1.6.2 there is a root system $\Phi(w_C)$ on $i\mathfrak{t}^{w_C}$ such that the image under orthogonal projection of $\widetilde{\Delta}_C^\vee$ is an extended set of simple coroots, and such that the extended coroot diagram is $\widetilde{D}^\vee/\langle C\rangle$. By Corollary 6.6.6, $\overline{\mathrm{g}}$ is a positive integral multiple of the coroot integer function on $\widetilde{\Delta}_C$. Let $\Phi(w_C, k) =$

$\Phi(it^{w_C}(\overline{\mathbf{g}},k))$. Theorem 1.7.4 now follows by applying Corollary 9.1.15 and Theorem 9.4.1 to the root system $\Phi(w_C)$, the function $\overline{\mathbf{g}}$, and the integer k.

CHAPTER 10

The Chern-Simons invariant

10.1. An algebraic invariant of c-triples

We introduce an invariant $CS_G(\mathbf{x})$ of a c-triple $\mathbf{x}$, which is a refinement of the order, and which we will relate to the Chern-Simons invariant of the corresponding flat bundle over the three-torus later in this chapter.

First let us record the following useful lemma concerning simple, non-simply laced groups (which of course also follows directly from classification).

LEMMA 10.1.1. *Let G be simple and non-simply laced. Then the order of $\mathcal{C}G$ is at most 2. If the order of $\mathcal{C}G$ is 2, then $\widetilde{D}(G)$ is either a chain with two multiple bonds at the ends or has one multiple bond meeting one leaf and one trivalent vertex which meets the remaining two leaves.*

PROOF. The Dynkin diagram $D(G)$ of a simple, non-simply laced group is a chain with a single multiple bond. Therefore, $\widetilde{D}(G)$ is either a chain with at most two multiple bonds or has one multiple bond and one trivalent vertex. The proof is then an elementary argument involving the possible diagram automorphisms of $\widetilde{D}(G)$. □

Let G be simple and let I_0^G be the unique Weyl invariant positive definite inner product on $i\mathfrak{t}$ with the property that $I_0^G(a^\vee, a^\vee) = 2$ for every short coroot a. The form I_0^G extends uniquely to a complex bilinear form on $\mathfrak{t} \otimes \mathbf{C}$ which is negative definite on $\mathfrak{t}$. It is easy to check that, for all roots a and all $t \in i\mathfrak{t}$,

$$I_0^G(a^\vee, t) = \frac{h_a}{g_a} a(t).$$

In particular, if a is a long root, then $I_0^G(a^\vee, t) = a(t)$. There is an induced inner product on the Lie algebra of any maximal torus of G, which we also denote by I_0^G. Now suppose that $G = \prod_{i=1}^r G_i$, where the G_i are simple. Set $\mathfrak{t}_i = \mathfrak{t} \cap \mathrm{Lie}(G_i)$ and define $I_0^G = \sum_i I_0^{G_i}$.

LEMMA 10.1.2. *Let G be simple and let $\widetilde{\Delta} = \widetilde{\Delta}(G)$. Let $\widetilde{I} \subset \widetilde{\Delta}$ be a proper subset and $H \subseteq G$ be the semi-simple subgroup whose complexified Lie algebra is generated by the root spaces $\mathfrak{g}^a$ for $\pm a \in \widetilde{I}$. Let $\widetilde{H} = \prod_i H_i$ be the decomposition of the universal covering $\widetilde{H}$ of H into simple factors, and $\widetilde{I} = \coprod_i I_i$ be the corresponding decomposition. Let $\mathfrak{t}_H = \mathfrak{t} \cap \mathrm{Lie}(H)$ and $\mathfrak{t}_i = \mathfrak{t} \cap \mathrm{Lie}(H_i)$. Then $\mathfrak{t}_H = \bigoplus_i \mathfrak{t}_i$ and $I_0^G | i\mathfrak{t}_H = \sum_i \epsilon_i I_0^{H_i}$, where d_i is the highest root of H_i with respect to I_i and $\epsilon_i = I_0^G(d_i^\vee, d_i^\vee)/2$. In particular, if G is simply laced or if H_i contains a root which is a long root of G, then $\epsilon_i = 1$.*

PROOF. The inner product $I_0^G | i\mathfrak{t}_i$ is invariant under the Weyl group of H_i, and thus is a multiple of $I_0^{H_i}$. Clearly, this multiple is $I_0^G(d_i^\vee, d_i^\vee)/2$. □

Let $\mathbf{x} = (x, y, z)$ be a c-triple such that the c-pair (x, y) is in normal form. Let $\widetilde{Z}(z)$ be the universal covering of $Z(z)$. Choose a maximal torus $T(z)$ for $Z(z)$. Of course, $T(z)$ is a maximal torus of G. Let $\mathfrak{t}(z)$ be its Lie algebra. Let $\tilde{x}$, resp. $\tilde{y}$, be a lift of x, resp. y to $\widetilde{Z}(z)$, and let $\tilde{c} = [\tilde{x}, \tilde{y}] \in \widetilde{Z}(z)$. Then $\tilde{c} \in \mathcal{C}\widetilde{Z}(z)$. Choose $\zeta \in i\mathfrak{t}(z)$ with $\exp(2\pi i\zeta) = \tilde{c} \in \widetilde{Z}(z)$. Since $z \in \mathcal{C}Z(z)$, there is an element $\hat{z} \in i\mathfrak{t}(z)$ with $\exp(2\pi i\hat{z}) = z$.

DEFINITION 10.1.3. We define $CS_G(\mathbf{x}) = [I_0^G(\zeta, \hat{z})] \in \mathbf{R}/\mathbf{Z}$. The *order* of $CS_G(\mathbf{x})$ is its order as an element of $\mathbf{R}/\mathbf{Z}$.

LEMMA 10.1.4. *The value of $CS_G(\mathbf{x})$ depends only on the conjugacy class of $\mathbf{x}$.*

PROOF. We begin by showing that $CS_G(\mathbf{x})$ only depends on $\mathbf{x}$ and not any of the choices made above. First fix the maximal torus $T(z)$. Then the choice of $\hat{z} \in i\mathfrak{t}(z)$ is determined up to an element $\lambda^\vee$ in the coroot lattice of G. Since $\exp(2\pi i\zeta)$ is an element of the center of G, we see that $I_0^G(\lambda^\vee, \zeta) \in \mathbf{Z}$. Thus $[I_0^G(\zeta, \hat{z})]$ is independent of the choice of $\hat{z}$. Now fix $\hat{z}$ and vary ζ. If $\exp(2\pi i\zeta') = \tilde{c}$, then $\zeta' - \zeta$ is an element of the coroot lattice of $\widetilde{Z}(z)$. Since $\exp(2\pi i\hat{z})$ is a central element of $Z(x)$, we see that $I_0^G(\zeta' - \zeta, \hat{z}) \in \mathbf{Z}$.

Finally consider another maximal torus $T'(z)$ in $Z(z)$. There is an element $g \in \widetilde{Z}(z)$ conjugating $T'(z)$ to $T(z)$. Since c and z are central in $Z(z)$, conjugation by g fixes $\tilde{c}$ and z. This establishes that $CS_G(\mathbf{x})$ is well-defined. Clearly, then, it is a conjugacy class invariant. □

LEMMA 10.1.5. *Let $\mathbf{x} = (x, y, z)$ and $\mathbf{x}' = (x', y', z')$ be c-triples in G which lie in the same component of the moduli space. Then $CS_G(\mathbf{x}) = CS_G(\mathbf{x}')$.*

PROOF. Choose a maximal torus $S \subseteq T$ of $Z^0(x, y, z)$. Choose a maximal torus $T(z)$ for $Z(z)$ with $S \subseteq T(z)$. Let $L = DZ(S)$. Then $c \in L$ and there is a rank zero c-triple $(x_0, y_0, z_0) = \mathbf{x}_0$ in L and elements $(s, t, u) \in S \times S \times S$ such that $(x, y, z) = (sx_0, ty_0, uz_0)$. It suffices by Theorem 2.3.1 to show that $CS_G(\mathbf{x}) = CS_G(\mathbf{x}_0)$. We fix a maximal torus T_L for L such that $z_0 \in T_L$. Clearly $Z_L(z_0) \subseteq Z_G(z)$. Let $\tilde{x}_0, \tilde{y}_0 \in \widetilde{Z}_L(z_0)$ be lifts of x_0, y_0, let $\tilde{c}_0 = [\tilde{x}_0, \tilde{y}_0]$, and let $\zeta_0 \in i\mathfrak{t}_L(z_0)$ be such that $\exp(2\pi i\zeta_0) = \tilde{c}_0$. If $\hat{z}_0 \in i\mathfrak{t}(z_0)$ satisfies $\exp(2\pi i\hat{z}_0) = z_0$, then $CS_G(x_0, y_0, z_0) = [I_0^G(\zeta_0, \hat{z}_0)]$. We can lift x to $\tilde{x} \in \tilde{S} \cdot \tilde{x}_0$, where $\tilde{S}$ is the identity component of the inverse image of S in $\widetilde{Z}(z)$, and similarly for $\tilde{y}$. Thus $[\tilde{x}, \tilde{y}] = [\tilde{x}_0, \tilde{y}_0] = \tilde{c}_0$ and hence $CS_G(x, y, z_0) = [I_0^G(\zeta_0, \hat{z}_0)] = CS_G(x_0, y_0, z_0)$.

Lastly, replace z_0 by $uz_0 \in Sz_0$. Clearly, there is an element $\hat{u} \in i\mathfrak{s} = i\mathrm{Lie}(S)$ with the property that $\exp(2\pi i\hat{u}) = u$. Thus, $\hat{u} + \hat{z}_0 \in i\mathfrak{t}(z)$ has the property that $\exp(2\pi i(\hat{u} + \hat{z}_0)) = uz_0$. Thus, $CS_G(\mathbf{x}) = [I_0^G(\zeta_0, \hat{u} + \hat{z}_0)]$. But $\mathfrak{s} = \mathfrak{t}_L^\perp$ under the pairing I_0^G, so that $I_0^G(\zeta_0, \hat{u}) = 0$. Hence $CS_G(\mathbf{x}) = CS_G(x, y, z_0) = CS_G(x_0, y_0, z_0)$. This completes the proof. □

Given a component X of the moduli space $\mathcal{T}_G(c)$ of c-triples in G, we define $CS_G(X)$ to be $CS_G(\mathbf{x})$, where $\mathbf{x}$ is any c-triple whose conjugacy class lies in X.

PROPOSITION 10.1.6. *Let (x, y) be a c-pair. Then the function $Z(x, y) \to \mathbf{R}/\mathbf{Z}$ defined by $z \mapsto CS_G(x, y, z)$ induces a homomorphism $\pi_0(Z(x, y)) \to \mathbf{R}/\mathbf{Z}$. Hence the order of $CS_G(\mathbf{x})$ divides the order of $\mathbf{x}$.*

PROOF. The fact that $z \mapsto CS_G(x, y, z)$ descends to a function $\pi_0(Z(x, y)) \to \mathbf{R}/\mathbf{Z}$ is immediate from Lemma 10.1.5. To see that it is a homomorphism, since $\pi_0(Z(x, y))$ is cyclic by Corollary 7.5.3, it suffices to show that $CS_G(x, y, z^\ell) = \ell CS_G(x, y, z)$ for all $z \in Z(x, y)$. Let $\tilde{x}, \tilde{y} \in \widetilde{Z}(z)$ be lifts of x, y, let $\tilde{c} = [\tilde{x}, \tilde{y}]$, and let $\zeta \in i\mathfrak{t}(z)$ satisfy $\exp(2\pi i \zeta) = \tilde{c}$. Finally, let $\tilde{z} \in i\mathfrak{t}(z)$ satisfy $\exp(2\pi i \tilde{z}) = z$. Then $CS_G(x, y, z) = I_0^G(\zeta, \tilde{z}) \bmod \mathbf{Z}$. Since $Z(z) \subseteq Z(z^\ell)$, for the element $\ell \tilde{z} \in i\mathfrak{t}(z) = i\mathfrak{t}(z^\ell)$ we have $\exp(2\pi i \ell \tilde{z}) = z^\ell$. Clearly, then, $CS_G(x, y, z^\ell) = [I_0^G(\zeta, \ell \tilde{z})] = \ell CS_G(x, y, z)$. □

We determine the order of CS_G in case $G = SU(n+1)$:

LEMMA 10.1.7. *Suppose that (x, y, z) is a c-triple in $SU(n+1)$, where c generates the center of $SU(n+1)$. Then the order of $CS_G(x, y, z)$ is the order of $z \in \mathcal{C}SU(n+1)$.*

PROOF. This follows easily from Lemma 9.4.3. □

10.1.1. More on the structure of L_c. The existence of the invariant CS_G leads to more detailed, classification-free information on the structure of L_c:

PROPOSITION 10.1.8. *Suppose that G is simple and let $c \in \mathcal{C}G$ be of order $o(c) > 1$. Let $n_0 = \gcd\{g_{\overline{a}}\}$. Then:*

(1) *$L_c = \prod_{i=1}^r L_i$ where each L_i is a simply connected, simple group of type A_{n_i} for some $n_i \geq 1$;*
(2) *$\operatorname{lcm}\{n_i + 1 : i = 1, \dots, r\} = o(c)$, and there is an i for which $n_i + 1 = o(c)$.*
(3) *If G is simply laced, then $o(c) = n_0$.*
(4) *If G is not simply laced, then either $n_0 = 1$, in which case each L_i is of type A_1 and the roots of L_i are short roots of G, or $n_0 = 2$, in which case each L_i is of type A_1, and exactly one of the L_i has a simple root which is a long root of G.*
(5) *If G is not simply laced and $n_0 = 1$, then $\Delta(c)$ contains the unique short simple root which is not perpendicular to at least one long simple root.*
(6) *If G is not simply laced and $n_0 = 2$, then $\Delta(c)$ contains a long simple root a, $\exp(2\pi i \varpi_a^\vee) = c$ and all other simple roots of G are short.*
(7) *There is a c-triple (x, y, t) in G of order n_0, where (x, y) is a c-pair in normal form and $t \in T^{w_c}$, such that the order of $CS_G(x, y, t)$ is also n_0.*

PROOF. By Theorem 3.6.1, we know that $L_c = \prod_i L_i$ is isomorphic to a product of simply connected groups of type A_{n_i} and that c is a product of elements $c_i \in \mathcal{C}L_i$ generating the center of L_i. It follows immediately that $o(c) = \operatorname{lcm}\{n_i + 1 : i = 1, \dots, r\}$. If G is of type A_N for some $N \geq 1$, then L_c is a product of simple groups of type A_{n-1} where $n = o(c)$, and so Parts 1 and 2 hold in this case. Assume that G is not of type A. Then $\widetilde{D}(G)$ is contractible and has at most two vertices of order > 2. Furthermore, if it has a vertex of order > 3, then it is $\widetilde{D}_4$. Any diagram automorphism of such a diagram has order $1, 2, 3$, or 4. Since the center acts faithfully on $\widetilde{D}(G)$, $o(c)$ is divisible by at most one prime. It follows that $n_i + 1 = o(c)$ for some i. This proves Parts 1 and 2.

We number the L_i so that $n_1 + 1 = o(c)$, and we let $\mathfrak{t}_i = \mathfrak{t} \cap \operatorname{Lie}(L_i)$. Suppose G is simply laced. Let $t \in \mathcal{C}H_1$ be an element of order $o(c)$ and let (x, y) be a c-pair in L_c. Clearly, $\mathcal{C}L_c \subseteq T^{w_c}$ and hence $t \in T^{w_c}$. Let $\tilde{t} \in i\mathfrak{t}_1$ satisfy $\exp(2\pi i \tilde{t}) = t$ and let $\zeta_1 \in i\mathfrak{t}_1$ satisfy $\exp(2\pi i \zeta_1) = c_1$. Since G is simply laced, it follows from

Lemma 10.1.2 that $CS_G(x,y,t) = CS_{L_c}(x,y,t) = [I_0^{H_1}(\tilde{t},\zeta_1)]$. Since t and c_1 each generate $\mathcal{C}H_1$, Lemma 10.1.7 shows that $[I_0^{H_1}(\tilde{t},\zeta_1)]$ is of order $n_1+1 = o(c)$. By Lemma 10.1.6, $o(c)$ divides the order of (x,y,t) in G. Since $t \in T^{w_c}$, the order of (x,y,t) in G divides n_0 by Proposition 8.2.9. If $\overline{a}$ is the orbit containing the extended root, then n_0 divides $g_{\overline{a}} = o(c)$. It follows that all of the above divisibilities are in fact equalities. In particular $o(c) = n_0$ and the order of $CS_G(x,y,t)$ is n_0. Thus we have proved Part 3, as well as Part 7 in the simply laced case.

To treat the non-simply laced case, we need the following.

CLAIM 10.1.9. *Let G be non-simply laced and let $c \in \mathcal{C}G, c \neq 1$. Let $\overline{I}_0^G(c,c) \in \mathbf{R}/\mathbf{Z}$ be defined as follows: Choose $\zeta, \mu \in i\mathfrak{t}$ with $\exp(2\pi i\zeta) = \exp(2\pi i\mu) = c$, and set $\overline{I}_0^G(c,c) = [I_0^G(\zeta,\mu)]$. Then $\overline{I}_0^G(c,c)$ is well-defined. Finally, $CS_G(\mathbf{x}) = 0$ for every c-triple $\mathbf{x}$ in L_c if and only if $\overline{I}_0^G(c,c) = 0$ if and only if $c \in S^{w_c}$ if and only if T^{w_c} is connected.*

PROOF. Since $\zeta, \mu \in P^\vee$, it follows immediately that varying ζ and μ by elements in $Q^\vee$ changes $I_0^G(\zeta,\mu)$ by an integer, showing that $\overline{I}_0^G(c,c)$ is well-defined.

By Lemma 10.1.1 $o(c) = 2$. Hence, by Part 2, all the L_i are of type A_1. Let $\{a_1,\ldots,a_t\}$ be the simple roots of L_c. Then a representative for $(2\pi i)^{-1}\log(c)$ is $\zeta_0 = \sum_i (1/2)a_i^\vee$. Another representative is $\mu_0 = \varpi_a^\vee$ for some simple root a of G. Since $\varpi_a^\vee$ represents an element of $\mathcal{C}G$, $h_a = 1$, and thus $g_a = h_a$ and a is a long root of G.

Let $\mathbf{x}_0 = (x,y,c)$ be a c-triple in L_c. By definition, for any representatives ζ and μ for $(2\pi i)^{-1}\log(c)$, we have $CS_G(\mathbf{x}_0) = [I_0^G(\zeta,\mu)]$. Thus $\overline{I}_0^G(c,c) = CS_G(\mathbf{x}_0)$. Hence, if $CS_G(\mathbf{x}) = 0$ for every c-triple $\mathbf{x}$ in L_c, then $\overline{I}_0^G(c,c) = 0$. By Lemma 10.1.2,

$$I_0^G(\zeta_0,\mu_0) = I_0^G\Big(\sum_i (1/2)a_i^\vee, \varpi_a^\vee\Big) = (1/2)\sum_i \epsilon_i\delta_{a,a_i}.$$

Thus $I_0^G(\zeta_0,\mu_0)$ is zero if a is distinct from all the a_i. Using the fact that a is a long root and hence $\epsilon_i = 1$ if $a = a_i$, it follows that $I_0^G(\zeta_0,\mu_0)$ is equal to $1/2$ if a is equal to one of the a_i. In particular, if $\overline{I}_0^G(c,c) = 0$, then $\varpi_a^\vee$ is orthogonal to all the $a_i^\vee$ and thus is orthogonal to $i\mathfrak{t}_{L_c}$. In this case, $\varpi_a^\vee \in i\mathfrak{t}^{w_c}$ and hence $c \in S^{w_c}$. Thus, $\overline{I}_0^G(c,c) = 0$ implies that $c \in S^{w_c}$.

By Lemma 10.1.1, $\mathcal{C}G = \langle c\rangle$. On the other hand, by Lemma 8.1.13, the map $\mathcal{C}G \to \pi_0(T^{w_c})$ is surjective. If $c \in S^{w_c}$, then $\pi_0(T^{w_c}) = 0$ and hence T^{w_c} is connected.

Finally suppose that T^{w_c} is connected. Then, by Corollary 7.5.3, $n_0 = 1$. Let $\mathbf{x}$ be a c-triple in L_c. Then by Proposition 8.2.9, the order of $\mathbf{x}$ divides n_0 and hence is 1. Since the order of $CS_G(\mathbf{x})$ divides the order of $\mathbf{x}$, by Proposition 10.1.6, it follows that $CS_G(\mathbf{x}) = 0$. □

Returning to the proof of the proposition, let us suppose that G is non-simply laced and that $o(c) = 2$, so that each L_i is of type A_1. Let $a_i \in \Delta$ be the simple root of L_i. Let $a \in \Delta$ be the unique simple root such that $c = \exp(2\pi i\varpi_a^\vee)$. Since $h_a = 1$, the root a is long. Let $\zeta \in \mathfrak{t}_{L_c}$ be such that $\exp(2\pi i\zeta) = c \in L_c$ and let $z_i = \exp(2\pi i(1/2)a_i^\vee)$. Then

$$CS_G(x,y,z_i) = [I_0^G((1/2)a_i^\vee,\zeta)] = [\epsilon_i a_i(\zeta)/2].$$

Since $\exp(2\pi i\zeta)$ projects to a generator of the center of L_i, we see that

$$CS_G(x,y,z_i) = [\epsilon_i/2].$$

On the other hand, using the lift $2\pi i\varpi_a^\vee$ for c shows that

$$CS_G(x,y,z_i) = [I_0^G((1/2)a_i^\vee, \varpi_a^\vee)] = [\epsilon_i\delta_{a_i,a}/2].$$

Since a is long, $\epsilon_i\delta_{a_i,a} = \delta_{a_i,a}$. It follows that $\epsilon_i \equiv \delta_{a_i,a} \bmod 2$. In other words, every a_i distinct from a is short. Hence either no root of L_c is a long root of G, in which case $\overline{I}_0^G(c,c) = 0$, or exactly one simple root of L_c is a long root of G and $\overline{I}_0^G(c,c) \neq 0$. In the first case, T^{w_c} is connected, and hence $n_0 = 1$, and in the second case T^{w_c} has two components and so $n_0 = 2$. The argument also shows that, if $a_i = a$, then the order of $CS_G(x,y,z_i)$ is exactly $2 = n_0$. Of course, if $n_0 = 1$, then the order of $CS_G(x,y,z_i)$ is also 1. This proves Part 4, as well as Part 7 in the non-simply laced case.

The proofs of Parts 5 and 6 are very similar, and we shall just prove Part 6. Suppose that G is not simply laced and that one of the a_i, say a_1, is a long root of G. Since the Dynkin diagram for G has a unique multiple bond, the long roots form a connected chain. Thus, if there is another long simple root $b \neq a_1$ of G, there is a long simple root b of G with $I_0^G(a_1^\vee, b^\vee) = -1$. Since the a_i are short roots for $i > 1$, and hence $\epsilon_i = 2$, we have that $I_0^G(a_i^\vee, b^\vee) = \epsilon_i a_i(b^\vee) \equiv 0 \pmod 2$ for all $i > 1$. It follows that $b(\sum_i (1/2)a_i^\vee) = I_0^G(b^\vee, \sum_i(1/2)a_i^\vee) \equiv (1/2) \bmod \mathbf{Z}$. This is impossible since $\exp(2\pi i \sum_i (1/2)a_i^\vee) = c \in \mathcal{C}G$. Thus a is the unique long simple root in G. □

The proof of Parts 5 and 6 actually shows the following: The simple roots in $\Delta(c)$ are given as follows. The Dynkin diagram of G is a chain, and the node at one end is a short root. Begin with this node, and then take every other node in the diagram until you reach the node of the double bond. Thus, for G of type B_n, $\Delta(c)$ is a single node corresponding to the short simple root, and for G of type C_n, $\Delta(c)$ consists of $n/2$ short simple roots if n is even and consists of $(n-1)/2$ short simple roots plus the long simple root if n is odd.

10.1.2. Order of $CS_G(\mathbf{x})$ in the rank zero case.

PROPOSITION 10.1.10. *Let G be simple and suppose that $\mathbf{x}$ is a c-triple of rank zero in G. Then the order of $CS_G(\mathbf{x})$ equals that of $\mathbf{x}$. For every k dividing exactly one of the $\{g_{\overline{a}} : \overline{a} \in \widetilde{\Delta}_c\}$, the function CS_G defines a bijection between the set of conjugacy classes of c-triples of rank zero and order k in G and the points of order k in $\mathbf{R}/\mathbf{Z}$.*

PROOF. Fix k dividing exactly one of the integers $g_{\overline{a}}$. By Proposition 8.1.7 and Proposition 10.1.6 it suffices to exhibit a single c-triple $\mathbf{x}$ such that $CS_G(\mathbf{x})$ has order $g_{\overline{a}}$.

Let (x,y,z) be a c-triple of rank zero and order $g_{\overline{a}}$. By Lemma 2.2.3, z is conjugate to the exponential of $(2\pi i)$ times a vertex of the alcove A. Moreover, $Z(z)$ is semi-simple and contains the rank zero c-pair (x,y). Thus by Proposition 4.1.1, the universal cover $\widetilde{Z}(z)$ is a product of groups H_i of type A_{n_i} and there is a lift $\tilde{c}$ of c to $\widetilde{Z}(z)$ such that the image of $\tilde{c}$ generates the center of every simple factor. Let $\mathfrak{t}(z) = \bigoplus_i \mathfrak{t}_i$ be the orthogonal direct sum decomposition induced by the decomposition of $\widetilde{Z}(z) = \prod_{i=1}^{r} H_i$. We write $\zeta = \sum_i \zeta_i$ where $\zeta_i \in i\mathfrak{t}_i$ and $2\pi i \zeta_i$ exponentiates to a generator of $\mathcal{C}H_i$.

Let $\tilde{z}$ be a lift of z to $\widetilde{Z}(z)$. Then $\tilde{z}$ lies in the center of $\widetilde{Z}(z)$. Hence, if m is the least common multiple of the integers $n_i + 1$, then $\tilde{z}^m = 1$. Since the image of z in the group $\pi_0(Z(x,y))$ has order $g_{\overline{a}}$, it follows that $g_{\overline{a}}|m$. For every $z' \in \mathcal{C}Z(z)$ the triple (x,y,z') is a c-triple. We shall find a $z' \in \mathcal{C}Z(z)$ such that $CS_G(x,y,z')$ has order m. Supposing this, by Corollary 10.1.6, it follows that m divides the order of (x,y,z'). Thus, $g_{\overline{a}}$ divides the order of (x,y,z'), which in turn divides at least one of the $g_{\overline{b}}$. It follows that $g_{\overline{b}} = g_{\overline{a}} = m$ and that the order of (x,y,z') is $g_{\overline{a}}$. Thus, the order of $CS_G(x,y,z')$ is equal to $g_{\overline{a}}$.

It remains to construct the required element $z' \in \mathcal{C}Z(z)$. For any z' in $\mathcal{C}Z(z)$, let $\hat{z}' \in i\mathfrak{t}(z)$ be such that $\exp(2\pi i \hat{z}') = z'$. We write $\hat{z}' = \sum_i \hat{z}_i'$ with $\hat{z}_i' \in i\mathfrak{t}_i$. Clearly,

$$CS_G(x,y,z') = [I_0^G(\zeta, \hat{z}')] = \sum_i \epsilon_i [I_0^{H_i}(\zeta_i, \hat{z}_i')],$$

Since $\exp(2\pi i \zeta_i)$ is a generator of the center of H_i and H_i is isomorphic to $SU(n_i + 1)$, it follows from Lemma 10.1.7 that, for every $r_i \in \mathbf{R}/\mathbf{Z}$ of order dividing $n_i + 1$, there is an element $\hat{z}_i' \in i\mathfrak{t}_i$, with $2\pi i \hat{z}_i'$ exponentiating to an element contained in the center of H_i, such that $[I_0^{H_i}(\zeta_i, \hat{z}_i')[= r_i \bmod \mathbf{Z}$. For appropriate choices of elements r_i of order dividing $n_i + 1$, the element $\sum_i r_i$ has order $m = \mathrm{lcm}\,\{n_i + 1 : i = 1, \ldots, r\}$. Consequently, there is an element $\hat{z}' = \sum_i \hat{z}_i'$ such that $\sum_i I_0^{H_i}(\zeta_i \hat{z}_i')$ is of order m modulo $\mathbf{Z}$.

If G is simply laced, then all the roots of H_i are long roots of G, and hence the factors ϵ_i are all one. In this case, the element $\hat{z}' \in i\mathfrak{t}(z)$ constructed in the last paragraph satisfies $z' = \exp(2\pi i \hat{z}')$ is in the center of $Z(z)$ and $I_0^G(\zeta, \hat{z}')$ is of order m modulo $\mathbf{Z}$.

Now suppose that G is non-simply laced and $c = 1$, so that $g_{\overline{a}} = g_a$. We write $Z(z) = \prod H_i/\langle \zeta \rangle$ where H_1 is the factor containing the highest root of G. Then, by Corollary 3.6.2, the factor H_1 is of type $A_{g_a - 1}$ and the image of $\pi_1(Z(z))$ in this factor is the center of A_{g_a-1}. Since one of the roots of this factor is the highest root of G, the roots of this factor are long roots of G. Choose a generator c of $\pi_1(Z(z))$, and let $c_i \in H_i$ be the image of c under the projection $\widetilde{Z}(z) \to H_i$. Since H_1 is of type A_{g_a-1} and c_1 generates the center of H_1, there is a c_1-triple $\mathbf{x}_1$ in H_1 of order g_a. By Lemma 10.1.7, $CS_{H_1}(\mathbf{x}_1)$ is also of order $g_{\overline{a}}$. For each $i > 1$ there is a c_i-triple $\mathbf{x_i}$ in H_i of order one. Clearly, the product $\prod_i \mathbf{x}_i$ is a $\prod c_i$-triple in $\widetilde{Z}(z)$, automatically of rank zero. It projects to a commuting triple $\mathbf{x}$ of rank zero in $Z(z)$. Clearly, $CS_G(\mathbf{x}) = \sum_i \epsilon_i CS_{H_i}(\mathbf{x}_i)$. Since the order of $\mathbf{x_i}$ is one for all $i > 1$, and since the roots of H_1 are long roots of G we see that $CS_G(\mathbf{x}) = CS_{H_1}(\mathbf{x}_1)$ and hence has order g_a. This proves the proposition in case $c = 1$ or G is simply laced.

There remains the possibility that G is not simply laced and $c \neq 1$. However, as the next lemma shows, there is just one possible G in this case. (Of course, this also follows easily from 8.1.5, which is based on classification.)

LEMMA 10.1.11. *Suppose that G is not simply laced and $c \neq 1$. If there is a rank zero c-triple (x,y,z) in G, then G is of type C_2.*

PROOF. By Lemma 10.1.1, $o(c) = 2$. Since $\widetilde{Z}(z)$ is a product of groups of type A, $z \neq 1$, and hence z is the exponential of $2\pi i$ times a vertex of the alcove A contained in a wall of A corresponding to the highest root. Let b be the simple root of G so that the face $\{b = 0\}$ of A is opposite to the vertex v with $\exp(2\pi iv) = z$. By Lemma 10.1.1, $\widetilde{D}(G)$ is either a chain with two multiple bonds at the ends or has one multiple bond meeting one leaf and one trivalent vertex, two of whose ears are the remaining two leaves. Moreover, the complement of the node corresponding to b is a diagram which is a union of diagrams of type A. It is easy to see that the only possibilities for such extended diagrams $\widetilde{D}(G)$ are $\widetilde{C}_2$, $\widetilde{B}_3$, or $\widetilde{B}_4$. The three possibilities for the quotient coroot integers are $1, 2$ in the case of $\widetilde{C}_2$, $1, 2, 2$ in the case of $\widetilde{B}_3$, and $1, 2, 2, 2$ in the case of $\widetilde{B}_4$. Thus, by Proposition 8.1.7, only in case $G = C_2$ does G contain a rank zero c-triple. □

Returning to the proof of Proposition 10.1.10 in the case where G is of type C_2, we let $\mathbf{x} = (x, y, z)$ be any rank zero c-triple. Up to conjugation, it follows that z is the exponential of $2\pi i$ times the vertex opposite the wall defined by $\{b = 0\}$, where b is the unique short simple root of C_2. Hence $Z(z) = \widetilde{Z}(z) \cong H_1 \times H_2$, where each H_i is of type A_1 and the extended root $-d$ is a simple root for one of the H_i, say H_1. Moreover $c = c_1c_2$, where the c_i is the nontrivial central element of H_i, and (x, y) is a product of rank zero c_i-pairs (x_i, y_i) in H_i. Furthermore z is the exponential of $(2\pi i)(1/2)d^\vee$. By Lemma 10.1.7, $CS_{H_1}(x_1, y_1, z) = 1/2$. Since $z \in H_1$, $CS_G(x, y, z) = CS_{H_1}(x_1, y_1, z) = 1/2$. □

10.1.3. The order of $CS_G(\mathbf{x})$.

PROPOSITION 10.1.12. *Let k be a positive integer dividing at least one of the $g_{\overline{a}}$ and suppose that $k \nmid n_0$. Let $S = S^{w_c}(k)$ and let $L = DZ(S)$. Then there exists a rank zero c-triple $\mathbf{x}$ in L of order k in G such that $CS_G(\mathbf{x})$ has order k.*

PROOF. We begin with the following lemma on the structure of L:

LEMMA 10.1.13. *Suppose that $c \neq 1$. With L as above, L_c is properly contained in L and L has a unique component L_0 which is not of A-type. Write $L = L_0 \times L'$. If G is not simply laced, then L_0 is of type C_2 and all simple factors of L' are of type A_1 whose roots are short roots of G.*

PROOF. By Proposition 8.2.9, since k does not divide n_0, L_c is properly contained in L and L has a unique simple factor L_0 not of A-type. Clearly if G is not simply laced, then L_0 is also not simply laced. In particular the two nodes of the double bond for G are simple roots for L_0. According to Proposition 10.1.8, one of these nodes lies in $\Delta(c)$. It follows that the projection c_0 of c to L_0 is nontrivial. Since L_0 contains a rank zero c_0-pair, it follows from Lemma 10.1.11 that L_0 is of type C_2. By Proposition 10.1.8, L_c is a product of groups of type A_1 and has at most one simple root which is a long root of G. If there is a long simple root in L_c, the corresponding node is a node of the double bond of $D(G)$. Hence it is a root of L_0. It follows that all of the simple factors of L' are of type A_1 whose roots are short roots of G. □

Returning to the proof of Proposition 10.1.12, first assume that G is simply laced. Let $L = \prod_{i=0}^r L_i$ be the product decomposition of L in simple factors, where L_0 is the factor which is not of type A. Let $c = \prod_i c_i$ be the corresponding decomposition of c. For each i, suppose that we are given a c_i-triple of rank zero $\mathbf{x}_i$ in L_i of order k_i. Then $\prod_i \mathbf{x}_i$ is a c-triple in L. Moreover, since G is simply laced, $CS_G(\mathbf{x}) = \sum_i CS_{L_i}(\mathbf{x}_i)$.

Let $\mathbf{x}$ be a c-triple in L whose order in G is k. Let $\mathbf{x}_i$ be the image of $\mathbf{x}$ in L_i and let k_i be its order as a c_i-triple in L_i. Note that $\mathbf{x}_i$ has rank zero in L_i. Thus by Proposition 10.1.10, the order of $CS_{L_i}(\mathbf{x}_i)$ is the order of $\mathbf{x}_i$ as a c_i-triple in L_i, and hence $CS_{L_i}(\mathbf{x}_i) = [r_i/k_i]$ for some integer r_i relatively prime to k_i. The order of $\mathbf{x}$, as a c-triple in L, is the least common multiple ℓ of the k_i. By Corollary 8.2.5, $k|\ell$. It is an elementary number-theoretic argument that there exist a_i such that $[r_0/k_0 + \sum_{i\geq 1} a_i/k_i] \in \mathbf{R}/\mathbf{Z}$ has order ℓ. For $i \geq 1$, L_i is of type A_{n_i}, and thus, given the integer a_i mod k_i, there exists a rank zero c_i-triple $\mathbf{x}_i'$ in L_i such that $CS_{L_i}(\mathbf{x}_i') = [a_i/k_i]$. Note that, by Proposition 10.1.10, the order of $\mathbf{x}_i'$ in L_i divides k_i. Replace $\mathbf{x}$ by the c-triple $\mathbf{x}' = \mathbf{x}_0 \cdot \prod_{i\geq 1} \mathbf{x}_i'$. Then $CS_G(\mathbf{x}') = \sum_{i\geq 0} CS_{L_i}(\mathbf{x}_i')$ is of order ℓ. On the other hand, the order of $\mathbf{x}'$ in L, which is the least common multiple of the orders of the $\mathbf{x}_i'$, divides ℓ. By Proposition 10.1.6, the order of $CS_G(\mathbf{x}')$ divides the order of $\mathbf{x}'$ as a c-triple in G, which in turn by Corollary 8.2.5 divides the order of $\mathbf{x}'$ as a c-triple in L which divides ℓ which is the order of $CS_G(\mathbf{x}')$. Therefore, the order of $\mathbf{x}'$ in G is the order of $CS_G(\mathbf{x}')$, namely ℓ. We write $\mathbf{x}' = (x', y', z')$. Then $\mathbf{x}'' = (x', y', (z')^{\ell/k})$ is a c-triple of order k in G such that the order of $CS_G(\mathbf{x}'')$ is also k.

Next suppose that $c = 1$. In this case $L = L_0$ is simple. By Claim 5.5.2, there is a simple root for L which is a long root of G. Choose a rank zero commuting triple $\mathbf{x}$ in L of order k in G. By Lemma 10.1.2 and Proposition 10.1.10, the order of $CS_G(\mathbf{x}) = CS_L(\mathbf{x})$ is k.

We may thus assume that G is not simply laced and that $c \neq 1$. It follows from Lemma 10.1.13 that L_0 is of type C_2 and all simple factors of L' are of type A_1 whose roots are short roots of G. Thus $k = 2$. Let $\mathbf{x} = \mathbf{x}_0 \cdot \mathbf{x}'$ be a c-triple in L. By Lemma 10.1.2, $CS_G(\mathbf{x}) = CS_{L_0}(\mathbf{x}_0)$. Choose a rank zero c_0-triple $\mathbf{x}_0$ in L_0. Then its order is 2 and by Proposition 10.1.10, the order of $CS_{L_0}(\mathbf{x}_0)$ is also 2. Choose any rank zero c'-triple $\mathbf{x}'$ in L'. Its order divides 2. Thus the order of $\mathbf{x} = \mathbf{x}_0 \cdot \mathbf{x}'$ in L is 2. Since the order of $CS_G(\mathbf{x})$ is 2, it follows that the order of $\mathbf{x}$ in G is also 2. This concludes the proof. □

THEOREM 10.1.14. *Let G be simple, let $c \in \mathcal{C}G$, and let $\mathbf{x}$ be a c-triple. Then the order of $\mathbf{x}$ is equal to the order of $CS_G(\mathbf{x})$. For any $k \geq 1$ dividing at least one of the $g_{\overline{a}}$, the function CS_G induces a bijection between the components of $\mathcal{T}_G(c)$ of order k and the points in $\mathbf{R}/\mathbf{Z}$ of order k.*

PROOF. Fix a k dividing at least one of the $g_{\overline{a}}$. There is the corresponding group L containing a rank zero c-triple of order k in G. It follows from Proposition 8.3.1 that there exists a rank zero c-triple (x_0, y_0, z_0) in L, such that every component X of $\mathcal{T}_G(c)$ of order k in G contains the conjugacy class of (x_0, y_0, z_0^ℓ) for exactly one ℓ where $1 \leq \ell \leq k$ and ℓ is relatively prime to k. By Proposition 10.1.6, $CS_G(x_0, y_0, z_0^\ell) = \ell CS_G(x_0, y_0, z_0)$. Since CS_G is constant on connected components, it clearly suffices to find, for every k, a c-triple $\mathbf{x}$ of order k in G such that the order of $CS_G(\mathbf{x})$ is also k. This follows from Part 7 of Proposition 10.1.8 and Proposition 10.1.12. □

10.2. Flat connections and the Chern-Simons invariant

10.2.1. Relation of C-triples and flat $G/\langle C\rangle$-connections. Let Γ be a flat connection on a principal G-bundle ξ over the three-torus $T^3 = S^1 \times S^1 \times S^1$. Since G is connected and simply connected, ξ is trivial. Given a trivialization of ξ, the holonomy of Γ around the three coordinate circles is a commuting triple (x, y, z) in G. Varying the trivialization conjugates (x, y, z). Thus, the isomorphism class of the G-bundle and flat connection determines the conjugacy class of the commuting triple. This sets up an isomorphism between $\mathcal{T}_G$ and the moduli space of isomorphism classes of flat connections on principal G-bundles over the three-torus.

If C is not the identity, then a C-triple (x, y, z) in G does not determine a flat connection on a principal G-bundle over the three-torus. It does determine a flat connection on a principal K-bundle over the three-torus, where $K = G/\langle C\rangle$, but the isomorphism class of this flat connection determines and is determined by the conjugacy class of the image commutative triple $(\overline{x}, \overline{y}, \overline{z})$ in K, which is not the same as the conjugacy class in G of (x, y, z). To deal with this incompatibility, we shall consider an enhanced notion of flat connections up to isomorphism.

Fix a compact connected group K. Denote its universal cover by $\widetilde{K}$. Let M be a manifold, let ξ be a principal K-bundle over M and let $X \subset M$ be a subspace with the homotopy type of a connected one-complex and which carries the fundamental group. A *lifting of ξ over X* is a pair $(\tilde{\xi}, f)$, where

(1) $\tilde{\xi}$ is a principal $\widetilde{K}$-bundle over X;
(2) $f\colon \tilde{\xi} \times_{\widetilde{K}} K \to \xi|X$ is an isomorphism of principal K-bundles.

An *enhanced K-bundle* $(\xi, \tilde{\xi}, f)$ over (M, X) consists of an underlying K-bundle ξ together with a lifting $(\tilde{\xi}, f)$ over X. An *isomorphism* between two enhanced K-bundles $(\xi, \tilde{\xi}, f)$ and $(\xi', \tilde{\xi}', f')$ consists of an underlying K-bundle isomorphism $\sigma\colon \xi \to \xi'$ and an isomorphism $\tilde{\sigma}\colon \tilde{\xi} \to \tilde{\xi}'$ of $\widetilde{K}$-bundles such that $(\sigma|X) \circ f = f' \circ (\tilde{\sigma} \times_{\widetilde{K}} \mathrm{Id}_K)$.

LEMMA 10.2.1. *Let ξ be a a principal K-bundle ξ over M and let $\tau\colon X \times K \to \xi|K$ be a trivialization over X. Then there is an enhanced K-bundle $\Xi_\tau = (\xi, X \times \widetilde{K}, f_\tau)$, where f_τ is the composition $(X \times \widetilde{K}) \times_{\widetilde{K}} K = X \times K \xrightarrow{\tau} \xi|K$.*

(1) *Given an enhanced K-bundle $\Xi = (\xi, \tilde{\xi}, f)$ over (M, X) there is a trivialization τ of $\xi|X$ such that Ξ_τ is isomorphic to Ξ by an isomorphism whose underlying K-bundle isomorphism is the identity.*
(2) *Given two trivializations τ and τ' of $\xi|X$, the enhanced K-bundles Ξ_τ and $\Xi_{\tau'}$ are isomorphic if and only if the function $\kappa\colon X \to K$ corresponding to the automorphism $\tau^{-1} \circ \tau'$ of the trivial bundle lifts to a function from X to $\widetilde{K}$.*

PROOF. Since $\widetilde{K}$ is connected, every principal $\widetilde{K}$-bundle over X is isomorphic to the trivial bundle. Part 1 follows. A lifting of the K-bundle isomorphism $\tau^{-1} \circ \tau'$ to an automorphism of the trivial $\widetilde{K}$-bundle $X \times \widetilde{K}$ is the same as a lifting of κ to $\widetilde{K}$. This proves Part 2. □

Suppose that $\xi \to M$ is K-bundle, that $\tau\colon X \times K \to \xi|K$ is a trivialization, and that Γ is a flat connection on ξ. Then the flat connection $\tau^*(\Gamma|X)$ lifts uniquely to a flat connection $\widetilde{\Gamma}$ on $X \times \widetilde{K}$. The holonomy of $\widetilde{\Gamma}$ is a homomorphism $\rho_{(\Gamma,\tau)}\colon \pi_1(X) \to \widetilde{K}$, called the *$\widetilde{K}$-holonomy* of (Γ, τ).

LEMMA 10.2.2. *Let ξ, resp. ξ' be a K-bundle over M with a flat connection Γ, resp. Γ', and let τ, resp. τ', be a trivialization of $\xi|X$, resp. $\xi'|X$. Suppose that there* *is an isomorphism* $(\sigma, \tilde{\sigma})$ *from* Ξ_τ *to* $\Xi'_{\tau'}$ *with* $\sigma^*\Gamma' = \Gamma$. *Then the* $\widetilde{K}$*-holonomies* $\rho_{(\Gamma,\tau)}$ *and* $\rho_{(\Gamma',\tau')}$ *are conjugate by an element of* $\widetilde{K}$.

PROOF. Let $\widehat{\sigma}\colon X \times K \to X \times K$ be the K-bundle map obtained from $\tilde{\sigma}$ by dividing out by $\pi_1(K) \subseteq \mathcal{C}\widetilde{K}$. The connection $\tilde{\sigma}^*\widetilde{\Gamma}'$ on $X \times \widetilde{K}$ lifts the connection $\widehat{\sigma}^* f_{\tau'}^*\Gamma$ on $X \times K$. Since $f_{\tau'} \circ \tilde{\sigma} = \sigma \circ f_\tau$ and since $\sigma^*\Gamma' = \Gamma$, it follows that $\widehat{\sigma}^* f_{\tau'}^*\Gamma' = f_\tau^*\Gamma$. Thus, $\tilde{\sigma}^*\widetilde{\Gamma}'$ lifts $f_\tau^*\Gamma$. But $\widetilde{\Gamma}$ is the unique lifting of $f_\tau^*\Gamma$. It follows that $\widetilde{\Gamma} = \tilde{\sigma}^*\widetilde{\Gamma}'$, and hence the holonomies of $\widetilde{\Gamma}$ and $\widetilde{\Gamma}'$ are conjugate in $\widetilde{K}$. □

Let (Ξ, Γ) be a pair consisting of an enhanced K-bundle over (M, X) and a flat connection on the underlying K-bundle. Two pairs (Ξ, Γ) and (Ξ', Γ') are *isomorphic* if there exists an isomorphism $(\sigma, \tilde{\sigma})$ from Ξ to Ξ' such that $\sigma^*\Gamma' = \Gamma$. We define the $\widetilde{K}$*-holonomy* of (Ξ, Γ) to be the conjugacy class of the homomorphism $\rho_{(\Gamma,\tau)}\colon \pi_1(X) \to \widetilde{K}$ where $\tau\colon X \times K \to \xi|K$ is any trivialization with the property that there is an isomorphism from Ξ_τ to Ξ which is the identity on the underlying K-bundles. According to Lemmas 10.2.1 and 10.2.2 the $\widetilde{K}$-holonomy of (Ξ, Γ) is well-defined. We have established the following.

PROPOSITION 10.2.3. *Let* M *be a manifold and* $X \subset M$ *a subset with the homotopy type of a one-complex carrying the fundamental group of* M. *The* $\widetilde{K}$*-holonomy of a pair* (Ξ, Γ) *consisting of an enhanced* K*-bundle over* (M, X) *and a flat connection* Γ *on the underlying* K*-bundle is a conjugacy class of homomorphisms from* $\pi_1(X)$ *to* $\widetilde{K}$. *This* $\widetilde{K}$*-holonomy depends only on the isomorphism class of the pair.*

In case when M is the N-torus $T^N = (S^1)^N$ and $X \subset T^N$ deformation retracts onto the union $\bigvee_N S^1$ of the coordinate circles, the $\widetilde{K}$-holonomy of a pair (Ξ, Γ) defines a conjugacy class of homomorphisms from $\pi_1(X)$ to $\widetilde{K}$. We identify this conjugacy class with a conjugacy class of N-tuples $(x_1, \ldots, x_N) \in (\widetilde{K})^N$, where x_i is the image of the element in $\pi_1(X)$ represented by the ith coordinate circle.

PROPOSITION 10.2.4. *Suppose that* G *is the universal cover of* K. *Let* $T^N = (S^1)^N$ *be the* N*-torus and let* $X \subset T^N$ *deformation retract onto the union* $\bigvee_N S^1$ *of the coordinate circles. Suppose that* Ξ *is an enhanced* K*-bundle on* (T^N, X) *with underlying bundle* ξ, *and suppose that* Γ *is a flat connection on* ξ. *Then the* G*-holonomy of* (Ξ, Γ) *defines a conjugacy class of almost commuting* N*-tuples* $(x_1, \ldots, x_N) \in G^N$. *If* $c_{ij} = [x_i, x_j]$, *then* $c_{ij} \in \pi_1(K) \subseteq \mathcal{C}G$ *and is equal to the* 2*-dimensional characteristic class of* ξ *evaluated on the* $(ij)^{\text{th}}$*-coordinate two-torus. The map which associates to the pair* (Ξ, Γ), *consisting of an enhanced* K*-bundle over* (T^N, X) *and a flat connection on the underlying* K*-bundle, its* G*-holonomy is a bijection from the set of isomorphism classes of such pairs* (Ξ, Γ) *to the space of conjugacy classes of almost commuting* N*-tuples in* G.

PROOF. Given a flat K-connection on the two-torus with holonomy $\overline{x}, \overline{y}$ around the coordinate circles, let $x, y \in G$ be lifts of these elements. Then the commutator $[x, y] \in G$ lies in $\pi_1(K) \subseteq \mathcal{C}G$ and is equal to the characteristic class $w(\xi) \in H^2(T^2; \pi_1(K))$. Applying this to the various coordinate two-tori inside the N-torus, shows that the N-tuple associated to (Ξ, Γ) proves the first statement. The rest of the proposition is a straightforward exercise. □

Next we show that two enhanced K-bundles over (T^3, X) whose underlying K-bundles are isomorphic are isomorphic as enhanced K-bundles.

LEMMA 10.2.5. *Let ξ be a K-bundle over T^3. Suppose that τ_1 and τ_2 are two trivializations of $\xi|X$. Then there is a bundle automorphism σ of ξ such that $\sigma^*\tau_1 = \tau_2$.*

PROOF. Since the question only involves the homotopy type of the pair (M, X), we may assume that $X = \bigvee_3 S^1$. Over X, we can take $\sigma = \tau_1 \circ \tau_2^{-1}$. Let $\kappa\colon X \to K$ be the corresponding map. Since $\xi|X$ is trivial, so is $\mathrm{Aut}(\xi)|X$, and hence the fundamental group of T^3 acts trivially on the fundamental group of the fiber of $\mathrm{Aut}(\xi)$, which is isomorphic to K. Thus the obstructions to extending σ to T^3 lie in $H^i(T^3, X; \pi_{i-1}(K))$. The only nonzero such group is $H^2(T^3, X; \pi_1(K))$. The value of the obstruction on a relative 2-cell e is $\kappa_*[\partial e] \in \pi_1(K)$. For the standard relative cell decomposition of (T^3, X) for which the 2-skeleton is the union of the T_{ij} and the corresponding cells are e_{ij}, the elements $[\partial e_{ij}]$ are commutators in $\pi_1(X)$. Since $\pi_1(K)$ is abelian, $\kappa_*[\partial e_{ij}]$ is trivial. Hence the automorphism extends. □

COROLLARY 10.2.6. *Let C be an anti-symmetric 3×3 matrix with values in $\pi_1(K) \subseteq \mathcal{C}G$. Let Ξ be an enhanced K-bundle over (T^3, X) with $C(\Xi) = C$. Then there is a bijection from the set of flat connections modulo automorphism of Ξ to $\mathcal{T}_G(C)$.*

10.2.2. The Chern-Simons invariant. Let G be simple and let $\hat{I}_0(t) = I_0^G(t,t)$.

LEMMA 10.2.7. *For all $t \in \mathfrak{t}$,*

$$\frac{1}{2}\hat{I}_0(t) = \frac{1}{2}I_0^G(t,t) = -\frac{1}{16\pi^2 g}\mathrm{Tr}(\mathrm{ad}(t)^2),$$

where g is the dual Coxeter number of G. Equivalently,

$$\hat{I}_0(s,t) = -\frac{1}{16\pi^2 g}\mathrm{Tr}(2\mathrm{ad}(s)\cdot \mathrm{ad}(t)).$$

PROOF. By a result of Looijenga [**14**] Lemma (1.2),

$$\sum_{a\in\Phi} a(t)^2 = 2gI_0(t,t).$$

On the other hand, the action of $\mathrm{ad}\, t$ on $\mathfrak{g}$ has eigenvalue $2\pi i a(t)$ on the root space $\mathfrak{g}^a$. The result is then a direct computation. □

The Weyl invariant quadratic form $\hat{I}_0$ on $\mathfrak{t}$ has a unique extension to an $\mathrm{ad}\, G$-invariant quadratic form on $\mathfrak{g}$, also denoted by $\hat{I}_0$. It follows from Chern-Weil theory that the quadratic form $\frac{1}{2}\hat{I}_0$ defines a cohomology class $\tilde{c}_2 \in H^4(BG; \mathbf{R})$. By [**15**], VI §6 Theorem 6.23 (see also [**3**]), $H^4(BG; \mathbf{Z}) \cong \mathbf{Z}$ and $\tilde{c}_2$ is an integral generator for $H^4(BG; \mathbf{Z})$. Let M be a manifold and let ξ be a principal G-bundle over M. There is a corresponding classifying map $p\ M \to BG$, and we set $c_2(\xi) = p^*\tilde{c}_2$. Given a connection A on ξ, the image of $c_2(\xi)$ in $H^4(M; \mathbf{R})$ is represented by the closed 4-form $\frac{1}{2}\hat{I}_0(F_A)$.

There is a secondary characteristic class associated to c_2, the *Chern-Simons invariant*. Let M be a three-manifold and $\xi \to M$ a principal G-bundle. Let Γ and A be connections on ξ. Then

$$\mathrm{CS}_\Gamma(A) = -\frac{1}{16\pi^2 g}\int_M \mathrm{Tr}\left(2a\wedge F_\Gamma + a\wedge d_\Gamma(a) + \frac{2}{3}a\wedge(a\wedge a)\right), \tag{10.1}$$

where $a = A - \Gamma \in \Omega^1(M; \mathrm{ad}\,\xi)$. Defined in this manner, CS_Γ is a real-valued function on the space of connections on a given bundle. The flat connections are the critical points of CS_Γ, and hence CS_Γ is constant on continuous path of flat connections.

Since CS_Γ is a secondary class associated to the primitive integral class c_2, if Γ and Γ' are gauge equivalent connections then $\mathrm{CS}_\Gamma(A) - \mathrm{CS}_{\Gamma'}(A)$ is an integer. As we vary over all connections Γ' gauge equivalent to Γ, we can vary the Chern-Simons invariant by an arbitrary integer. Similarly, if A and A_1 are gauge equivalent connections then $\mathrm{CS}_\Gamma(A) - \mathrm{CS}_\Gamma(A_1)$ is an integer, and as we vary A_1 over all connections gauge equivalent to A we can vary CS_Γ by an arbitrary integer. In other words, CS_Γ is a well-defined function from the space of gauge equivalence classes of connections into $\mathbf{R}/\mathbf{Z}$.

More generally, suppose that K is a compact group whose universal cover is G and that $\xi \to M$ is a principal K-bundle with connection A. The form $\frac{1}{2}\hat{I}_0(F_A)$ is a closed 4-form representing a characteristic class of ξ, which we will also denote by $c_2(\xi)$. The only difference is that this class need not be integral. In fact, it lies in $(1/n)\mathbf{Z}$, where the image of $H^4(BK; \mathbf{Z})$ in $H^4(BG; \mathbf{Z})$ is generated by $n\tilde{c}_2$. Given K-connections Γ and Γ' on ξ, we can still define $\mathrm{CS}_\Gamma(\Gamma') \in \mathbf{R}$ by Equation 10.1. As before, this is a secondary characteristic class associated to the four-dimensional class c_2 of K-bundles. The difference is that, in general, if Γ'' is a K-connection gauge equivalent to Γ', then

$$\mathrm{CS}_\Gamma(\Gamma') - \mathrm{CS}_\Gamma(\Gamma'') \in \frac{1}{n}\mathbf{Z},$$

and hence CS_Γ is only well-defined modulo $\frac{1}{n}\mathbf{Z}$ for isomorphism classes of K-connections. As the next lemma shows, CS_Γ is well-defined modulo $\mathbf{Z}$ on enhanced isomorphism classes of flat K-connections.

LEMMA 10.2.8. *Let Ξ be an enhanced K-bundle over (M, X), where M is a closed, oriented three-manifold, and suppose that there is an automorphism of Ξ such that the induced automorphism of the underlying K-bundle ξ is σ. Then*

$$\mathrm{CS}_\Gamma(\Gamma') - \mathrm{CS}_\Gamma(\sigma^*\Gamma') \in \mathbf{Z}.$$

PROOF. Let $\hat{\xi} \to M \times S^1$ be the K-bundle obtained from $\xi \times I \to M \times I$ by gluing the ends together by σ. The difference $\mathrm{CS}_\Gamma(\Gamma') - \mathrm{CS}_\Gamma(\sigma^*\Gamma')$ is equal to $\int_{M\times S^1} c_2(\hat{\xi})$. We choose a trivialization τ of $\xi|X$ such that there is an isomorphism Ξ_τ to Ξ whose underlying K-bundle isomorphism is the identity. Then in this trivialization $\sigma|X$ is given by a continuous map $\kappa\colon X \to K$. The fact that σ comes from an automorphism of the enhanced bundle means that κ lifts to a map $X \to G$. This means that $\hat{\xi}|X \times S^1$ lifts to a G-bundle, and hence, since G is connected and simply connected, $\hat{\xi}|X \times S^1$ is trivial. A standard obstruction theory argument then shows that $\hat{\xi}$ is isomorphic as a K-bundle to the connected sum of the product bundle $\xi \times S^1 \to M \times S^1$ and a K-bundle $\eta \to S^4$. Thus, $\int_{M\times S^1} c_2(\hat{\xi}) = \int_{S^4} c_2(\eta)$. But since S^4 is simply connected, $\eta \to S^4$ lifts to a G-bundle $\tilde{\eta}$ and thus $c_2(\eta) = c_2(\tilde{\eta})$ takes an integral value on S^4. □

COROLLARY 10.2.9. *Fix an enhanced K-bundle Ξ over (M, X), where M is a closed, oriented three-manifold, and a flat connection Γ_1 on Ξ. Then the function CS_{Γ_1} induces a well-defined function from the set of isomorphism classes of pairs (Ξ, Γ), where Γ is a flat connection on Ξ, to $\mathbf{R}/\mathbf{Z}$.*

One important property of the Chern-Simons invariant is the following additivity property:

LEMMA 10.2.10. *Let Γ_0, Γ_1 and A be connections on a K-bundle ξ over a three-manifold. Then*

$$\mathrm{CS}_{\Gamma_0}(A) = \mathrm{CS}_{\Gamma_1}(A) + \mathrm{CS}_{\Gamma_0}(\Gamma_1).$$

COROLLARY 10.2.11. *Let $\{\Gamma_t\}_{t\in[0,1]}$ and $\{A_t\}_{t\in[0,1]}$ be two continuous paths of flat connections. Then $\mathrm{CS}_{\Gamma_0}(A_0) = \mathrm{CS}_{\Gamma_1}(A_1)$.*

PROOF. Since the critical points of the functional $A \mapsto \mathrm{CS}_{\Gamma_0}(A)$ are the flat connections, $\mathrm{CS}_{\Gamma_0}(A_0) = \mathrm{CS}_{\Gamma_0}(A_1)$. In particular, $\mathrm{CS}_{\Gamma_0}(\Gamma_1) = 0$. The corollary now follows from the additivity formula of the previous lemma. □

10.3. The basic computation

Let (x, y, z) be a c-triple in a simply connected group G. Then $Z(z)$ is a connected group containing x, y and c. We denote by $\overline{Z}(z)$ the quotient $Z(z)/\langle c\rangle$ and we denote by $\widetilde{Z}(z)$ the universal covering of $Z(z)$. Let $\tilde{x}, \tilde{y}$ be lifts of x, y to the universal covering $\widetilde{Z}(z)$. Let $\zeta = [\tilde{x}, \tilde{y}]$ and let $\tilde{\zeta} \in i\mathfrak{z}(z)$ be an element with $\exp(2\pi i\tilde{\zeta}) = \zeta$. Let T^2 be the torus $\mathbf{R}^2/\mathbf{Z}^2$. Let $D \subseteq T^2$ be a closed disk and let T_0 be the closure of $T^2 - D$. Let ∂ be the boundary of T_0, with coordinate θ, $0 \le \theta \le 1$.

LEMMA 10.3.1. *With notation as above, there is a flat connection $\tilde{A}_0$ on the trivial $\widetilde{Z}(z)$-bundle over T_0 such that the holonomy of $\tilde{A}_0$ along the two coordinate circles in T_0 is given by $(\tilde{x}, \tilde{y})$, and $\tilde{A}_0|\partial = \tilde{\zeta}\, d\theta$.*

PROOF. Clearly, there exists a flat connection A_0' with the required holonomy on the bundle $T_0 \times \widetilde{Z}(z)$. The connections $A_0'|\partial$ and $\tilde{\zeta}\, d\theta$ have the same holonomy. It is then easy to see that there is an automorphism σ of the trivial bundle, supported near ∂, such that $\tilde{A}_0 = \sigma^* A_0'$ satisfies the conclusions of the lemma. □

There is an induced flat $\overline{Z}(z)$-connection $\overline{A}_0$ on $\xi_0 = T_0 \times \overline{Z}(z)$, whose holonomy along ∂ is the identity. Thus there is a trivialization τ_∂ of $\xi_0|\partial$ for which $\overline{A}_0|\partial$ is the product connection, and hence we can extend ξ_0 and the flat connection $\overline{A}_0$ to a $\overline{Z}(z)$-bundle ξ' with a flat connection A_0 over T^2. By construction, there is a trivialization τ_D of $\xi'|D$ extending τ_∂ and with respect to which $\overline{A}_0|D$ is a trivial connection. We denote by τ_0 the given trivialization of $\xi'|T_0 = \xi_0$.

LEMMA 10.3.2. *The composition $(\tau_0|\partial) \circ (\tau_D^{-1}|\partial)$ is the map $S^1 \times \overline{Z}(z) \to S^1 \times \overline{Z}(z)$ given by $(\theta, g) \mapsto (\theta, \zeta^\theta \cdot g) = (\theta, \exp(2\pi i\theta\tilde{\zeta}) \cdot g)$.*

PROOF. The composition $(\tau_0|\partial) \circ (\tau_D^{-1}|\partial)$ pulls back the trivial connection to $\tilde{\zeta} d\theta$. The composition is given by left multiplication by $\gamma\colon S^1 \to \overline{Z}(z)$, and moreover $\gamma^{-1}d\gamma = \tilde{\zeta} d\theta$. By changing the trivialization τ_D by a constant change of gauge, we can arrange that $\gamma(0) = 1$. It follows that $\gamma(\theta) = \exp(2\pi i\theta\tilde{\zeta}) = \zeta^\theta$. □

Now consider $T^3 = T^2 \times S^1$ where $S^1 = \mathbf{R}/\mathbf{Z}$ has coordinate u. The trivialization $\tau_0 \times S^1$ restricts to a trivialization τ of $\xi \times S^1$ over $X = T_0 \bigvee S^1 \subset T_0 \times S^1$.

There is a maximal torus T of $Z(z)$, and hence of G, with $\mathfrak{t} = \mathrm{Lie}(T)$, such that $\tilde{\zeta} \in i\mathfrak{t}$. Of course, T also contains the central element z of $Z(z)$. Let $\hat{z}$ be an element of it such that $\exp(2\pi i\hat{z}) = z$. For $u \in \mathbf{R}$, define $z^u = \exp(2\pi iu\hat{z})$. For all $u, \theta \in \mathbf{R}$, the elements z^u and ζ^θ lie in T and hence commute.

Define a connection A on $\xi' \times S^1$ over $T^2 \times S^1$ as follows. Over $D \times S^1$, A has connection 1-form $\hat{z}\,du$ with respect to the trivialization $\tau_D \times \mathrm{Id}_{S^1}$. Over $T_0 \times S^1$, A has connection 1-form $z^{-u}\tau_0^* A_0 z^u + \hat{z}du$ with respect to the trivialization $\tau_0 \times S^1$. Direct computation shows that the connections given on $D \times S^1$ and on $T_0 \times S^1$ are flat. Since z^u and ζ^θ commute for all u and θ, it follows that these two partial connections glue together to define a connection A on $\xi' \times S^1$. Clearly, A is flat, the restriction of A to $\xi' \times \{0\}$ is isomorphic to A_0, and the $\widetilde{Z}(z)$ holonomy of (A, τ) around the three circle factors gives the commuting triple $(\tilde{x}, \tilde{y}, \exp(2\pi i\hat{z}))$ in $\widetilde{Z}(z)$.

Let us define the K-bundle $\xi = \xi' \times_{\overline{Z}(z)} K$ over T^2. The connection A induces a flat connection on $\xi \times S^1$, which we continue to denote by A. The trivialization τ of $\xi' \times S^1|X$ induces a trivialization of $\xi \times S^1|X$ which we will also denote by τ. The G-holonomy of (A, τ) around the three coordinate circles is (x, y, z).

LEMMA 10.3.3. *Fix a connection Γ on $\xi' \times S^1$ with the following properties:*

(1) *If $p\colon \xi' \times S^1 \to \xi'$ is the natural projection, then $\Gamma = p^*\Gamma_0$ for some connection Γ_0 on ξ'.*
(2) *The restriction of Γ_0 to $\xi_0 = \xi'|T_0$ is trivial in the trivialization τ_0.*
(3) *The restriction of Γ_0 to $\xi'|D$ is a T-connection in the trivialization τ_D.*

Also denote by Γ the resulting K-connection on $\xi \times S^1$. Then

$$\mathrm{CS}_\Gamma(A) = I_0^G(\hat{z}, \tilde{\zeta}).$$

PROOF. Let A' be the connection on $\xi \times S^1$ which is trivial over $D \times S^1$ in the trivialization $\tau_D \times S^1$ and which is given by the one-form $z^{-u}A_0 z^u$ over $T_0 \times S^1$ in the trivialization $\tau_0 \times S^1$. As before, one sees easily that these two descriptions match over $\partial D \times S^1$.

Let us compute $\mathrm{CS}_{A'}(A)$. First notice that $a = A - A' = \hat{z}du$ over both $D \times S^1$ and $T_0 \times S^1$. Clearly, then $a \wedge a = 0$. Over $D \times S^1$ the connection A' is trivial. Thus, on this patch $d_{A'}(a) = F_{A'} = 0$, and consequently the Chern-Simons integrand vanishes on this patch. Over $T_0 \times S^1$ we have $d_{A'}(a) = [z^{-u}A_0z^u, \hat{z}du]$ so that $a \wedge d_{A'}(a) = 0$. Lastly, on this patch $F_{A'} = du \wedge [z^{-u}A_0 z^u, \hat{z}]$ so that $a \wedge F_{A'} = 0$. This shows that $\mathrm{CS}_{A'}(A) = 0$ and hence, by Lemma 10.2.10, $\mathrm{CS}_\Gamma(A) = \mathrm{CS}_\Gamma(A')$.

Now we compute $\mathrm{CS}_\Gamma(A')$. First let us show that the Chern-Simons integrand for this invariant is zero over $D \times S^1$. Over $D \times S^1$, and with the trivialization $\tau_D \times S^1$, Γ is the pullback of a connection B on $\xi|D$ and A' is trivial. Thus $a = A' - \Gamma = -B$, F_B and $d_B(a)$ are all pulled back from forms on D. Thus, we see that over $D \times S^1$ the Chern-Simons integrand vanishes identically.

Next we compute the Chern-Simons integral over $T_0 \times S^1$ using the trivialization $\tau_0 \times S^1$. The connection Γ is trivial on this patch and the one-form a is $z^{-u}\tau_0^* A_0 z^u$. It follows that $d_\Gamma(a) = du \wedge [z^{-u}\tau_0^* A_0 z^u, \hat{z}] + z^{-u}\tau_0^* dA_0 z^u$. Thus, $\mathrm{CS}_\Gamma(A')$ is given by

$$\begin{aligned}
& -\frac{1}{16\pi^2 g}\int_{T_0\times S^1} \mathrm{Tr}\left(z^{-u}\tau_0^* A_0 z^u \wedge (du \wedge [z^{-u}\tau_0^* A_0 z^u, \hat{z}] + z^{-u}\tau_0^* dA_0 z^u)\right) \\
= \; & -\frac{1}{16\pi^2 g}\int_{T_0\times S^1} \mathrm{Tr}\left(z^{-u}\tau_0^* A_0 z^u \wedge (du \wedge [z^{-u}\tau_0^* A_0 z^u, \hat{z}])\right) \\
= \; & -\frac{1}{16\pi^2 g}\int_{T_0\times S^1} \mathrm{Tr}\left(-2(z^{-u}\tau_0^* A_0 z^u) \wedge (z^{-u}\tau_0^* A_0 z^u) \wedge \hat{z}du\right).
\end{aligned}$$

Since $z^{-u}\hat{z}z^u = \hat{z}$, we can rewrite this as

$$\mathrm{CS}_\Gamma(A') = -\frac{1}{16\pi^2 g}\int_{T_0\times S^1} \mathrm{Tr}\left(-2\tau_0^* A_0 \wedge \tau_0^* A_0 \wedge \hat{z}du\right).$$

Doing the u-integration and using the fact that $dA_0 + A_0 \wedge A_0 = 0$, we get

$$\begin{aligned}\mathrm{CS}_\Gamma(A') &= -\frac{1}{16\pi^2 g}\int_{T_0} \mathrm{Tr}\left(-2(\tau_0^* A_0 \wedge \tau_0^* A_0)\cdot \hat{z}\right)\\ &= -\frac{1}{16\pi^2 g}\int_{T_0} \mathrm{Tr}\left(2\tau_0^* dA_0 \cdot \hat{z}\right)\\ &= -\frac{1}{16\pi^2 g}\int_{\partial T_0} \mathrm{Tr}\left(2\tau_0^* A_0 \cdot \hat{z}\right).\end{aligned}$$

Since $\tau_0^* A_0|\partial T_0 = \tilde{\zeta}d\theta$, and using Lemma 10.2.7, we have

$$\mathrm{CS}_\Gamma(A) = \mathrm{CS}_\Gamma(A') = -\frac{1}{16\pi^2 g}\mathrm{Tr}(2\tilde{\zeta}\cdot\hat{z}) = [I_0(\hat{z},\tilde{\zeta})] = CS_G(x,y,z).$$

□

PROPOSITION 10.3.4. *Let Θ be a connection on $\xi \times S^1$ which is pulled back from a connection on ξ via the natural projection mapping. Then*

$$\mathrm{CS}_\Theta(A) = I_0^G(\hat{z},\tilde{\zeta}).$$

PROOF. By Lemma 10.2.10 it suffices to show that if Θ and Θ' are connections on $(\xi \times_{\overline{Z}(z)} K)\times S^1$ pulled back from connections on $(\xi \times_{\overline{Z}(z)} K)$ then $\mathrm{CS}_\Theta(\Theta') = 0$. But this is clear – under this hypothesis the Chern-Simons integrand (which is a three-form) is pulled back from a form on the two-torus T^2 and hence vanishes identically. □

COROLLARY 10.3.5. *Let $T^3 = T^2\times S^1$ and let $X \subset T^3$ be $T_0 \vee S^1$. Given $c \in \mathcal{C}G$ let $K = G/\langle c\rangle$. Fix a K-bundle $\xi_c \to T^2$ with $w(\xi_c) = c \in H^2(T^2;\pi_1(K)) \subseteq \mathcal{C}G$. Fix a trivialization τ_0 of $\xi_c|T_0$ and let τ_c be the trivialization of $(\xi_c\times S^1)|X$ obtained from τ_0 and the given product structure on the last S^1-direction. Let Ξ_c be the corresponding enhanced K-bundle over (T^3, X).*

(1) *For every c-triple $\mathbf{x}$, there is a flat connection $A_0(\mathbf{x})$ on $\xi_c \times S^1$ such that the G-holonomy of $(A_0(\mathbf{x}),\tau_c)$ is the conjugacy class of $\mathbf{x}$.*
(2) *For every flat connection A on $\xi_c\times S^1$ such that the G-holonomy of (A,τ_c) is the conjugacy class of $\mathbf{x}$, and for every connection Θ on $\xi_c \times S^1$ which is pulled back by the natural projection from a connection Θ_0 on ξ_c,*

$$\mathrm{CS}_\Theta(A) \equiv CS_G(x,y,z) \bmod \mathbf{Z}.$$

PROOF. The above construction shows that, given a c-triple $\mathbf{x}$, there is a K-bundle $\xi \to T^2$, a trivialization τ_0' of $\xi|T_0$ and a flat connection $A'(\mathbf{x})$ on $\xi \times S^1$ such that the G-holonomy of $A'(\mathbf{x})$ measured using the trivialization τ', which is the union of the trivialization τ_0' on T_0 with the product trivialization around the third coordinate circle, is $\mathbf{x}$. Since (x,y) is a c-pair, there is an isomorphism $\psi\colon \xi_c \to \xi$. By Lemma 10.2.5, there is a bundle isomorphism from $\xi_c \times S^1$ to $\xi \times S^1$ which carries the trivialization τ' to τ.

Let $A_0(\mathbf{x})$ be the pullback of the connection $A'(\mathbf{x})$. Then $A_0(\mathbf{x})$ is a flat connection on $\xi_c \times S^1$ whose G-holonomy is the conjugacy class of $\mathbf{x}$, and hence the G-holonomy of $(\Xi_c, A_0(\mathbf{x}))$ is the conjugacy class of $\mathbf{x}$.

By Proposition 10.3.4 $\mathrm{CS}_\Theta(A_0(x,y,z)) \equiv CS_G(x,y,z) \bmod \mathbf{Z}$. More generally, suppose that A is a flat connection on $\xi_c \times S^1$ such that the G-holonomy of $(\Xi_c, A(x,y,z))$ is the conjugacy class of $\mathbf{x}$. Then by Lemma 10.2.4 there is an enhanced automorphism of Ξ_c carrying A to $A_0(x,y,z)$. By Lemma 10.2.8 this implies that

$$\mathrm{CS}_\Gamma(A) - \mathrm{CS}_\Gamma(A_0(x,y,z)) \in \mathbf{Z},$$

and the result follows. □

Notice that the enhancement over $T_0 \subset T^2$ is irrelevant – it is only the enhancement in the last S^1-direction that is important. This is the connection analogue of the fact that if $\mathbf{x} = (x,y,z)$ is a c-triple, then multiplying x and y by powers of c does not change the G-conjugacy class of $\mathbf{x}$ and hence does not change its CS_G-invariant, but multiplying z by a power of c will change the conjugacy class of $\mathbf{x}$ in general, and will even change the connected component of $\mathcal{T}_G(c)$ containing the conjugacy class of $\mathbf{x}$, or equivalently will change $CS_G(\mathbf{x})$.

The case of commuting triples is worth stating separately.

PROPOSITION 10.3.6. *Let (x,y,z) be a commuting triple, and let $A(x,y,z)$ be a flat connection on a principal G-bundle ξ over T^3 with holonomy around the three coordinate circles equal to the conjugacy class of (x,y,z). Then ξ is a trivial bundle. Let Θ be a connection on this bundle isomorphic to the trivial connection. Then*

$$\mathrm{CS}_\Theta(A) \cong CS_G(x,y,z) \pmod{\mathbf{Z}}.$$

PROOF. Since $c = 1$, the G-bundle ξ_c given in the statement of Corollary 10.3.5 is trivial. Furthermore, an enhancement of a G-bundle is no extra information. The result is now immediate from this corollary and Lemma 10.2.8. □

Let $c \in \mathcal{C}G$ be given and let $K = G/\langle c\rangle$. We are now ready to define the Chern-Simons invariant of an isomorphism class of a pair (Ξ, A) consisting of an enhanced K-bundle over (T^3, X) whose underlying K-bundle is isomorphic to $\xi_c \times S^1$ and a flat connection. The G-holonomy of such a pair is the conjugacy class of a c-triple, and thus by Corollary 10.3.5 there is an enhanced isomorphism from Ξ_c to Ξ. Let σ be the underlying K-bundle isomorphism. Then we define

$$\mathrm{CS}(\Xi, A) = [\mathrm{CS}_\Theta(\sigma^* A)] \pmod{\mathbf{Z}},$$

where Θ is any flat connection on $\xi_c \times S^1$ pulled back from a flat connection on ξ_c. Note that, if Θ is such a connection, then its G-holonomy is the conjugacy class of $(x,y,1)$, where (x,y) is a c-pair, and, in particular, the G-holonomy lies in the component X_1 of c-triples of order 1. If $c = 1$, then we can take Θ to be the trivial connection and are computing the usual Chern-Simons invariant.

10.4. Proof of Theorem 1.8.1 and Theorem 1.8.2 in the case where $\langle C\rangle$ is cyclic

By Lemma 10.2.11, the value of $\mathrm{CS}(\Xi, A)$ only depends on the component X of $\mathcal{T}_G(c)$ containing the G-holonomy of (Ξ, A). Given this, the proof of Theorem 1.8.1 for c-triples is immediate from Corollary 10.3.5 and Theorem 10.1.14. The proof of Theorem 1.8.2 in the cyclic case follows from the dimension statement in Theorem 1.5.1, the computation of the Chern-Simons invariant contained in Theorem 10.1.14 and Corollary 10.3.5, and the numerology of Theorem 3.8.7, via the natural identification of $\mathbf{Z}/2g\mathbf{Z}$ with $\frac{1}{2g}\mathbf{Z}/\mathbf{Z}$.

CHAPTER 11

The case when $\langle C \rangle$ is not cyclic

In this chapter, we assume that $G = Spin(4n)$ for some $n \geq 2$. Then $\mathcal{C}G \cong \mathbf{Z}/2\mathbf{Z} \times \mathbf{Z}/2\mathbf{Z}$. Choose an identification of a maximal torus T for G with the quotient of $\mathbf{R}^{2n}$ with basis $\{e_i\}$ by the even integral lattice so that the roots for G are $(2\pi i)^{-1}(\{\pm e_i \pm e_j\})$. A set of simple roots for G is then $\Delta = \{a_1, \ldots, a_{2n-2}, a_-, a_+\}$, where $a_i = (2\pi i)^{-1}(e_i - e_{i+1})$ for $1 \leq i \leq 2n-2$, $a_- = (2\pi i)^{-1}(e_{2n-1} - e_{2n})$, and $a_+ = (2\pi i)^{-1}(e_{2n-1} + e_{2n})$. Label the non-trivial elements of $\mathcal{C}G$ as c_0, c_1, c_2 so that c_0 is represented by e_{2n}, c_1 is represented by $(\sum_i e_i)/2$, and c_2 is represented by $(e_1 + \cdots + e_{2n-1} - e_{2n})/2$. In terms of the simple roots, $c_0 \equiv (2\pi i)(\frac{1}{2}a_+ + \frac{1}{2}a_-) \bmod 2\pi i Q^\vee$, $c_1 \equiv (2\pi i)(\sum_{i=1}^{n-1} \frac{1}{2}a_{2i-1} + \frac{1}{2}a_+) \bmod 2\pi i Q^\vee$, and $c_2 \equiv (2\pi i)(\sum_{i=1}^{n-1} \frac{1}{2}a_{2i-1} + \frac{1}{2}a_-) \bmod 2\pi i Q^\vee$. Thus $\Delta(c_0) = \{a_+, a_-\}$ and $L_{c_0} = L_+ \times L_-$, where $L_\pm$ is a group of type A_1 whose simple root is $a_\pm$. Let $\Delta(\mathcal{C})$ be the subset of Δ consisting of all roots $a \in \Delta$ such that ϖ_a does not annihilate $\mathcal{C}G$. Thus

$$\Delta(\mathcal{C}) = \Delta(c_0) \cup \Delta(c_1) \cup \Delta(c_2) = \{a_1, a_3, \ldots, a_{2n-3}, a_-, a_+\}.$$

Then $L_{\mathcal{C}} = L_{\Delta(\mathcal{C})} = \prod_{i=1}^{n-1} L_i \times L_+ \times L_-$ is a product of $n+1$ groups of type A_1, where a_{2i-1} is the simple root of L_i. We also let $S_{\mathcal{C}} = S_{\Delta(\mathcal{C})}$.

We have the following analogue of Corollary 3.5.2, whose proof is left to the reader:

LEMMA 11.0.1. *Let $I \subseteq \Delta$ and let $L = L_I$. Then $L_{\mathcal{C}} \subseteq L$ if and only if $\mathcal{C} \subset L$, and $L = L_{\mathcal{C}}$ if and only if*

(1) *$\mathcal{C} \subset L$;*
(2) *L is a product of simple factors $L_i \cong SU(n_i)$ for some n_i;*
(3) *The projection of $\mathcal{C}$ to L_i generates $\mathcal{C}L_i$.*

With our choice of T and Δ, the action of w_{c_0} on $\mathfrak{t}$ is given by:

$$w_{c_0}(t_1, t_2, \ldots, t_{2n-1}, t_{2n}) = (-t_1, t_2, \ldots, t_{2n-1}, -t_{2n}),$$

so that $\mathfrak{t}^{w_{c_0}} = \{(t_1, t_2, \ldots, t_{2n-1}, t_{2n}) \in \mathfrak{t} : t_1 = t_{2n} = 0\}$. Of course, $\mathfrak{t}^{w_{c_0}}$ is conjugate to

$$\mathfrak{t}_{c_0} = \mathrm{Ker}(a_+) \cap \mathrm{Ker}(a_-) = \{(t_1, t_2, \ldots, t_{2n-1}, t_{2n}) \in \mathfrak{t} : t_{2n-1} = t_{2n} = 0\}.$$

Note that $c_0 \in S^{w_{c_0}}$ but that $c_1, c_2 \notin S^{w_{c_0}}$.

We shall consider triples (x, y, z) in G with $[x, y] = c_0$, $[x, z] = c_1$, and $[y, z] = c_2$, and let C be the corresponding antisymmetric matrix.

11.1. Rank zero C-triples

LEMMA 11.1.1. *There is a rank zero C-triple in G if and only if $G = Spin(8)$. For $G = Spin(8)$ there are exactly two conjugacy classes of rank zero C-triples. If*

(x, y, z) is a rank zero C-triple in Spin(8), then (x, y, z^{-1}) is a rank zero C-triple not conjugate to (x, y, z).

PROOF. Suppose that (x, y, z) is a rank zero C-triple in G. After conjugation, we may assume that (x, y) is a c_0-pair in normal form. Thus $S^{w_{c_0}}$ is a maximal torus of $Z^0(x, y)$. The element z commutes with x and y up to an element of the center. Since the fixed subgroup of conjugation by z on $Z(x, y)$ has rank zero, it follows from [**17**] II §2 that $Z^0(x, y)$ is a torus and hence $Z^0(x, y) = S^{w_{c_0}}$. By inspection, all of the integers $g_{\overline{a}}$, for the action of w_{c_0} on the coroot diagram of G are 2. Hence by Proposition 7.5.5, $Z(x, y) = T^{w_{c_0}}$. The triple (x, y, z^2) is a c_0-triple in G. By Proposition 8.2.9, $S^{w_{c_0}}$ is a maximal torus of $Z(x, y, z^2)$. This implies that z^2 acts trivially on $S^{w_{c_0}}$. Since z acts on this torus without fixed points, it follows that z acts by -1 on $\mathfrak{t}^{w_{c_0}}$, and hence $zsz^{-1} = s^{-1}$ for all $s \in S^{w_{c_0}}$.

By Proposition 3.5.4, the torus $S^{w_{c_0}}$ is conjugate to S_{c_0}. It will be convenient to conjugate x, y, z so that $Z^0(x, y) = S_{c_0}$. Of course, in this case $zsz^{-1} = s^{-1}$ for all $s \in S_{c_0}$. Write $x = s \cdot x_0$ and $y = s' \cdot y_0$ for elements $s, s' \in S_{c_0}$ and $x_0, y_0 \in L_{c_0}$. Since $c_1 s x_0 = {}^z(s x_0) = (s^{-1})^z x_0$, it follows that $s^2 = c_1({}^z x_0 x_0^{-1}) = c_1 \cdot u$ for some $u \in S_{c_0} \cap L_{c_0}$. This implies that s^2 is the image of an element of the form $(\frac{1}{2}, \frac{1}{2}, \dots, \frac{1}{2}, 0, 0)$ mod $\mathbf{Z}^n$. The same computation for y shows that $(s')^2$ is also the image of such an element. Thus, both s and s' are the images of elements of the form $(\pm\frac{1}{4}, \pm\frac{1}{4}, \dots, \pm\frac{1}{4}, 0, 0)$ mod $\mathbf{Z}^n$. It follows easily that, for $4n > 8$, there is a root of $Spin(4n)$ annihilating L_{c_0} and annihilating s and s'. This root then annihilates both x and y, contradicting the fact that $Z^0(x, y)$ is a torus. Hence $G = Spin(8)$.

Now let us describe all conjugacy classes of rank zero C-triples in $Spin(8)$. Let (x, y, z) be a rank zero C-triple in $Spin(8)$. As above, we arrange that $(x, y) = (s_1 x_0, s_2 y_0)$ with $s_i \in S_{c_0}$ and $(x_0, y_0) \in L_{c_0}$. Then $Z^0(x, y) = S_{c_0}$. According to Corollary 4.2.2, the conjugacy class of (x, y) depends only on the the pair $(\overline{s}_1, \overline{s}_2) \in \overline{S}_{c_0} = S_{c_0}/S_{c_0} \cap L_{c_0}$. It is easy to see that $\overline{S}_{c_0}$ is the quotient of $\mathbf{R}^2 \times \{0\} \times \{0\}$ by $\mathbf{Z}^2$. The image in the Weyl group of S_{c_0} of elements in $N_G(S_{c_0})$ which act trivially on L_{c_0} is $\mathbf{Z}/2\mathbf{Z} \times \mathbf{Z}/2\mathbf{Z}$, where one of the factors acts by -1 on $\mathfrak{t}_{c_0}$ and the other acts by switching the coordinates. The argument above shows that $\overline{s}_i$ is the image of an element of the form $(\pm\frac{1}{4}, \pm\frac{1}{4}, 0, 0)$ mod $\mathbf{Z}^2$. Let $\tilde{s}_i$ be a lift of $\overline{s}_i$ to $\mathfrak{t}_{c_0}$.

The element z normalizes S_{c_0} and acts by -1 on $\mathfrak{t}_{c_0}$. Thus z sends $S_{c_0} \cdot x$ to $S_{c_0} \cdot c_1 x$. Since $S_{c_0} \cdot x$ contains a regular element of T, the element z normalizes T and hence normalizes the unique maximal torus of L_{c_0} containing x_0. Since x_0 is the image of $(0, 0, \frac{1}{2}, 0)$, it follows that $z x_0 z^{-1}$ is the image of one of $(0, 0, \pm\frac{1}{2}, 0)$ or $(0, 0, 0, \pm\frac{1}{2})$. The fact that $zxz^{-1} = xc_1$ implies that $z x_0 z^{-1}$ is the image of $(0, 0, 0, \pm\frac{1}{2})$. By conjugation by an element commuting with x we can arrange that $z x_0 z^{-1}$ is the image of $(0, 0, 0, -\frac{1}{2})$. A computation shows that $\tilde{s}_1$ must be the image of an element congruent to $\pm(\frac{1}{4}, \frac{1}{4}, 0, 0)$ mod $\mathbf{Z}^2$. After conjugating by an element of the normalizer of S_{c_0}, we may assume that $\tilde{s}_1$ is the image of an element congruent to $(\frac{1}{4}, \frac{1}{4}, 0, 0)$ mod $\mathbf{Z}^2$. Since there is no root of $Spin(8)$ annihilating both s_1 and s_2, this means that $\tilde{s}_2$ is either $(\frac{1}{4}, -\frac{1}{4}, 0, 0)$ or $(-\frac{1}{4}, \frac{1}{4}, 0, 0)$ mod $\mathbf{Z}^2$. After conjugation by an element of the normalizer of S_{c_0} which interchanges the two coordinates, and hence fixes $\tilde{s}_1$, we may assume that $\tilde{s}_2$ is congruent to $(\frac{1}{4}, -\frac{1}{4}, 0, 0)$ mod $\mathbf{Z}^2$. This proves that, given a rank zero C-triple (x, y, z) in $Spin(8)$, the c_0-pair (x, y) is determined up to conjugation in $Spin(8)$. The representatives that we have chosen are: x is the image of $(\frac{1}{4}, \frac{1}{4}, \frac{1}{2}, 0))$ and y is the image of $(\frac{1}{4}, -\frac{1}{4}, 0, 0)) \cdot w$ where

w is an element of L_{c_0} normalizing the unique maximal torus of L_{c_0} containing $x_0 = \exp(0, 0, \frac{1}{2}, 0)$, and the image of w in the Weyl group of L_{c_0} is the product of the non-trivial elements in each factor of L_{c_0}.

Next we show that there is $z \in G$ such that (x, y, z) is a rank zero C-triple. Write $x_0 = x_0^+ x_0^-$, where $x_0^\pm \in L_\pm$. Similarly write $y_0 = y_0^+ y_0^-$. Let $z = \epsilon \cdot x_0^+ y_0^+$ where $\epsilon \in N_{Spin(8)}(S_{c_0})$ commutes with L_{c_0} and represents the Weyl element which is multiplication by -1 on $\mathfrak{t}_{c_0}$. Direct computation shows that (x, y, z) is a rank zero C-triple.

Suppose that (x, y, z) is a rank zero C-triple in $Spin(8)$. Any other rank zero C-triple is conjugate to (x, y, z'), where $z' = zg$ for some $g \in Z(x, y)$. Since $\pi_0(Z(x, y))$ is cyclic of order two and since $i_z = -\mathrm{Id}$ on $Z^0(x, y)$, there are exactly two conjugacy classes of rank zero C-triples in $Spin(8)$, one of which is represented by (x, y, z) and the other by (x, y, zt), where $t \in Z(x, y)$ but $t \notin Z^0(x, y)$. Direct computation shows that z is of order 4 and that z^2 is in the non-trivial component of $Z^0(x, y)$. Thus, $(x, y, z^{\pm 1})$ represent the two conjugacy classes of rank zero C-triples. □

11.2. Action of the center and of the outer automorphism group

The following lemma is easy and its proof is left to the reader.

LEMMA 11.2.1. *An automorphism of $Spin(8)$ which acts trivially on $\mathcal{C}Spin(8)$ is an inner automorphism and hence acts trivially on the space of of conjugacy classes of rank zero C-triples in $Spin(8)$.*

Let us consider the action of $(\mathcal{C}Spin(8))^3$ on the space of rank zero C-triples.

LEMMA 11.2.2. *Let $\mu \in \mathcal{C}Spin(8)$ act on the moduli space of conjugacy classes of rank zero C-triples in $Spin(8)$ by: $\mu \cdot (x, y, z) = (x, y, \mu \cdot z)$. Then this action is transitive. The isotropy subgroup of this action is $\langle c_0 \rangle$. Likewise, the action defined by $\mu \cdot (x, y, z) = (\mu \cdot x, y, z)$ is transitive, and its isotropy is $\langle c_2 \rangle$, and the action defined by $\mu \cdot (x, y, z) = (x, \mu \cdot y, z)$ is transitive, with isotropy $\langle c_1 \rangle$.*

PROOF. This is immediate from the explicit description given above. □

11.3. The general case

Following the general pattern, we begin by determining the possibilities for the maximal torus of $Z(x, y, z)$. Recall from Theorem 2.3.1 that, for every $\mathcal{C}G$-triple $(x, y, z) = \mathbf{x}$, there is a subset $I(\mathbf{x}) \subset \Delta$ such that S_I is conjugate in G to a maximal torus of $Z(x, y, z)$.

LEMMA 11.3.1. *Let $I' = \Delta(\mathcal{C}) \cup \{a_{2n-2}\}$. If $(x, y, z) = \mathbf{x}$ is a $\mathcal{C}G$ triple, then either $I(\mathbf{x}) = \Delta(\mathcal{C})$ or $I(\mathbf{x}) = I'$.*

PROOF. Let $L = L_{I(\mathbf{x})}$. Then by Lemma 11.0.1, since $\mathcal{C}G \subset L$, $\Delta(\mathcal{C}) \subseteq I(\mathbf{x})$. We may write $L = L_0 \times \prod_{i=1}^r L_i$, where each L_i is of type A and L_0 is either trivial or of type D_{2k} for some $k \geq 2$. For $0 \leq i \leq r$, let π_i denote projection to the ith factor. The projection of $\mathbf{x}$ to L_i is a rank zero $\pi_i(C)$-triple. Thus, if $i \geq 1$, then L_i is of type A_1. If L_0 is not trivial, then $\pi_0(\mathcal{C})$ is the full center of L_0. By Lemma 11.1.1, $k = 2$. Since $\Delta(\mathcal{C}) = \{a_1, a_3, \ldots, a_{2n-1}, a_-, a_+\}$, it is clear that the only possibilities for $I(\mathbf{x})$ which satisfy the above conditions are either $I(\mathbf{x}) = I_{\mathcal{C}}$ or $I(\mathbf{x}) = I'$. □

The tori $S_{\mathcal{C}}$ and $S_{I'}$ corresponding to the sets $\Delta(\mathcal{C})$ and I' have the following Lie algebras:

$$\begin{aligned}\mathfrak{t}_{\mathcal{C}} &= \bigcap_{i=1}^{n-1}\operatorname{Ker}(a_{2i-1})\cap\operatorname{Ker}(a_+)\cap\operatorname{Ker}(a_-)\\ &= \{(t_1,t_1,t_3,t_3,\ldots,t_{2n-3},t_{2n-3},0,0)\};\\ \mathfrak{t}_{I'} &= \bigcap_{i=1}^{n-1}\operatorname{Ker}(a_{2i-1})\cap\operatorname{Ker}(a_+)\cap\operatorname{Ker}(a_-)\cap\operatorname{Ker}(a_{2n-2})\\ &= \{(t_1,t_1,t_3,t_3,\ldots,t_{2n-5},t_{2n-5},0,0,0,0)\}.\end{aligned}$$

In the quotient diagram $\widetilde{D}^\vee(G)/\mathcal{C}G$, one of the quotient coroot integers $g_{\overline{a}}$ is 2 and all the others are 4. Thus, in terms of the quotient diagram and the quotient coroot integers, we have:

$$\mathfrak{t}^{w_C} = \{(0,s_2,\ldots,s_{n-1},s_n,-s_n,-s_{n-1},\ldots,-s_2,0)\}; \tag{11.1}$$

$$\mathfrak{t}^{w_C}(\overline{\mathbf{g}},4) = \{(0,s_2,\ldots,s_{n-1},0,0,-s_{n-1},\ldots,-s_2,0)\}. \tag{11.2}$$

From this, it is easy to establish the following:

LEMMA 11.3.2. *The torus $S_{\mathcal{C}}$ is conjugate to $S^{w_C} = S^{w_C}(\overline{\mathbf{g}},1) = S^{w_C}(\overline{\mathbf{g}},2)$. The torus $S_{I'}$ is conjugate to $S^{w_C}(\overline{\mathbf{g}},4)$.*

Having described the possible maximal tori, we need to determine how many components of the moduli space correspond to a given maximal torus. We begin with the following elementary computation.

LEMMA 11.3.3. *Let γ be the non-trivial central element in $SU(2)$. Consider triples (p,q,r) in $SU(2)$ with $[p,q]=1$ and $[p,r]=[q,r]=\gamma$. Then, up to conjugation, there are exactly two such triples. We have $p^2=q^2=r^2=\gamma$. The element pq^{-1} lies in $\mathcal{C}SU(2)$ and is a complete invariant of the conjugacy class of (p,q,r). The action of $(\mathcal{C}SU(2))^3$ on the conjugacy classes of such triples given by $(a,b,c)\cdot(p,q,r)=(ap,bq,cr)$ is transitive and the stabilizer of any conjugacy class is $\{(a,b,c): a=b\}$.*

PROOF. The pair (p,r) is a γ-pair in $SU(2)$. Also, $pq^{-1}\in Z(p,r)$ and hence lies in $\mathcal{C}SU(2)$. The result follows from Proposition 4.1.1 and Corollary 4.1.2. $\square$

Suppose that $\mathbf{x}$ is a C-triple in $L_{\mathcal{C}} = L_{\mathcal{C}} = \prod_{i=1}^{n-1}L_i\times L_+\times L_-$. The triple $\mathbf{x}$ is a product of triples $\prod_i \mathbf{x}_i\cdot\mathbf{x}_+\cdot\mathbf{x}_-$ in the factors. We write $\mathbf{x}_i=(x_i,y_i,z_i)$ and $\mathbf{x}_\pm=(x_\pm,y_\pm,z_\pm)$. Let $\epsilon_i\colon \mathcal{C}L_i\to\{\pm1\}$, $\epsilon_\pm\colon\mathcal{C}L_\pm\to\{\pm1\}$ be the unique isomorphisms. Then define $\epsilon(\mathbf{x})=\prod_{i=1}^{n-1}\epsilon_i(x_iy_i^{-1})\epsilon_+(x_+z_+^{-1})\epsilon_-(y_-z_-^{-1})$.

LEMMA 11.3.4. *There are exactly two components of $\mathcal{T}_G(C)$ with maximal torus conjugate to $S_{\mathcal{C}}$. Two C-triples $\mathbf{x}$ and $\mathbf{x}'$ in $L_{\mathcal{C}}$ lie in the same component of $\mathcal{T}_G(C)$ if and only if $\epsilon(\mathbf{x})=\epsilon(\mathbf{x}')$.*

PROOF. By Theorem 2.3.1, the number of components of $\mathcal{T}_G(C)$ with maximal torus conjugate to $S_{\mathcal{C}}$ is given by the number of conjugacy classes of rank zero C-triples in $L_{\mathcal{C}}$ modulo the action of $F_{\mathcal{C}}^3$ and of $W(S_{\mathcal{C}},G)$, where $F_{\mathcal{C}}=S_{\mathcal{C}}\cap L_{\mathcal{C}}I\subseteq \mathcal{C}L_{\mathcal{C}}\cong(\mathbf{Z}/2\mathbf{Z})^{n+1}$. Let us first consider the action of $F_{\mathcal{C}}^3$. It is easy to see that

$$F_{\mathcal{C}}=\{(\mu_1,\ldots,\mu_{n-1},\mu_+,\mu_-):\mu_1\cdots\mu_{n-1}=\mu_+=\mu_-\}. \tag{11.3}$$

Straightforward computation shows that the action of F_C^3 on the space of conjugacy classes of rank zero C-triples in L_C preserves the invariant ϵ and acts transitively on the set of conjugacy classes with a given ϵ.

The image of $W(S_C, G)$ in the outer automorphism group of $L_C = \prod_i L_i \times L_+ \times L_-$ is easily checked to be the subgroup of all permutations of the L_i factors. Thus the action of $W(S_C, G)$ fixes ϵ. This completes the proof of the lemma. □

LEMMA 11.3.5. *Under the action of F_C^3 on the space of conjugacy classes of rank zero C-triples in L_C the stabilizer of any class is the set of*

$$\boldsymbol{\mu} = (\boldsymbol{\mu}_1, \ldots, \boldsymbol{\mu}_{n-1}, \boldsymbol{\mu}_+, \boldsymbol{\mu}_-) \in F_C^3$$

such that

$$\mu_j^{(1)} = \mu_j^{(2)}, 1 \leq j \leq n-1, \mu_-^{(2)} = \mu_-^{(3)}, \mu_+^{(1)} = \mu_+^{(3)}, \tag{11.4}$$

where $\boldsymbol{\mu}_j = (\mu_j^{(1)}, \mu_j^{(2)}, \mu_j^{(3)})$ and similarly for $\boldsymbol{\mu}_\pm$.

PROOF. This is immediate from Lemma 11.3.3. □

COROLLARY 11.3.6. *There are two components of $\mathcal{T}_G(C)$ associated with the torus S_C, and each is homeomorphic to*

$$((S_C \times S_C \times S_C)/F)\,/W(S_C, G),$$

where $F \subset (\mathcal{C}L_C)^3$ is the set of $\boldsymbol{\mu}$ satisfying Equations 11.3 and 11.4.

Now let us consider the case when $I = I'$. Recall that $L_{I'} = \prod_{i=0}^{n-2} L_i$, where L_i is of type A_1 for $i > 0$ and $L_0 \cong Spin(8)$. If $\mathbf{x}$ is a rank zero C-triple in $L_{I'}$, then we write $\mathbf{x} = \prod_{i=0}^{n-2} \mathbf{x}_i$. The group L_i has two rank zero $\pi_i(C)$-triples, up to conjugation, where π_i is the projection from L to L_i. It follows that $L_{I'}$ has 2^{n-1} rank zero C-triples up to conjugation. If $\mathbf{x}_i$ and $\mathbf{x}_i'$ are two rank zero $\pi_i(C)$-triples in L_i, define $\delta_i(\mathbf{x}_i, \mathbf{x}_i') = 1$ if $\mathbf{x}_i$ is conjugate in L_i to $\mathbf{x}_i'$ and -1 otherwise. If $i > 0$, then it is easy to see that $\delta_i(\mathbf{x}_i, \mathbf{x}_i') = \epsilon_i(x_i y_i^{-1})\epsilon_i(x_i'(y_i')^{-1})$, in the notation introduced prior to Lemma 11.3.4. Define

$$\delta(\mathbf{x}, \mathbf{x}') = \prod_{i=0}^{n-2} \delta_i(\mathbf{x}_i, \mathbf{x}_i').$$

LEMMA 11.3.7. *Two rank zero C-triples $\mathbf{x}$ and $\mathbf{x}'$ in $L_{I'}$ lie in the same component of $\mathcal{T}_G(C)$ if and only if $\delta(\mathbf{x}, \mathbf{x}') = 1$. Hence there are exactly two components of $\mathcal{T}_G(C)$ with maximal torus conjugate to $S_{I'}$.*

PROOF. By Theorem 2.3.1, the number of components of $\mathcal{T}_G(C)$ with maximal torus conjugate to $S_{I'}$ is given by the number of conjugacy classes of rank zero C-triples in $L_{I'}$ modulo the action of $F_{I'}^3$ and of $W(S_{I'}, G)$, where $F_{I'} = S_{I'} \cap L_{I'} \subseteq \mathcal{C}L_{I'} \cong (\mathbf{Z}/2\mathbf{Z})^{n-2} \times \mathcal{C}L_0$. Let us first consider the action of $F_{I'}^3$. Let $\nu\colon \{\pm 1\} \to \mathcal{C}L_0$ be the embedding which sends -1 to $\pi_0(c_0)$. Then it is easy to check that

$$F_{I'} = \left\{ (\mu_0, \ldots, \mu_{n-2}) : \mu_0 = \nu\left(\prod_{i=1}^{n-2} \epsilon_i(\mu_i)\right) \right\} \subseteq \prod_{i=0}^{n-2} \mathcal{C}L_i. \tag{11.5}$$

Note that, by Lemma 11.2.2, if $\boldsymbol{\mu}$ is an element of $F_{I'}^3$ and $\mathbf{x}$ is a rank zero C-triple in $L_{I'}$, then $\delta(\mathbf{x}, \boldsymbol{\mu} \cdot \mathbf{x}) = 1$. Conversely, it is clear that, if $\delta(\mathbf{x}, \mathbf{x}') = 1$, then $\mathbf{x}$ and $\mathbf{x}'$ are in the same $F_{I'}^3$-orbit. Finally, the image of $W(S_{I'}, G)$ in the outer automorphism group of $L_{I'}$ is the permutation group of the factors L_i for $i \geq 1$. Hence $\delta(\mathbf{x}, w\cdot\mathbf{x}) = 1$ for all $w \in W(S_{I'}, G)$. The lemma follows. □

LEMMA 11.3.8. *Under the action of $F_{I'}^3$ on the space of conjugacy classes of rank zero C-triples in $L_{I'}$ the stabilizer of a class is the set of $\boldsymbol{\mu} = (\boldsymbol{\mu}_0, \ldots, \boldsymbol{\mu}_{n-2}) \in F_{I'}^3$ such that*

$$\mu_j^{(1)} = \mu_j^{(2)}, 1 \leq j \leq n-2, \tag{11.6}$$

where $\boldsymbol{\mu}_j = (\mu_j^{(1)}, \mu_j^{(2)}, \mu_j^{(3)})$.

PROOF. This is immediate from Lemma 11.3.3 and Lemma 11.2.2. □

COROLLARY 11.3.9. *There are two components of $\mathcal{T}_G(C)$ associated with the torus $S_{I'}$, and each is homeomorphic to*

$$((S_{I'} \times S_{I'} \times S_{I'})/F')\,/W(S_{I'}, G),$$

where $F' \subset (\mathcal{C}L_{I'})^3$ is the subgroup of elements satisfying Equations 11.5 and 11.6.

To each component X of $\mathcal{T}_G(C)$, we assign a positive integer which we shall call the *order* of X, in the following way. Suppose that X is a component containing the conjugacy class of $\mathbf{x}$, where $\mathbf{x}$ is a rank zero C-triple in $L_{\mathcal{C}}$. Then the order of X is 1 if $\epsilon(\mathbf{x}) = 1$ and the order of X is 2 otherwise. Each of the remaining components is associated to $S_{I'}$, and we define each of them to have order 4. Thus, for every positive integer k dividing 4, the number of components of $\mathcal{T}_G(C)$ is $\varphi(k)$.

Parts 1,2,3,4 of Theorem 1.5.1, in the case where $\langle C\rangle$ is not cyclic, are now contained in the statements of Lemma 11.3.1, Corollaries 11.3.6 and 11.3.9, and the definition of the order.

Lastly we prove Part 5 of Theorem 1.5.1 in the non-cyclic case. The explicit coordinates on the Lie algebras $\mathfrak{t}^{w_C}$ and $\mathfrak{t}^{w_C}(\overline{\mathfrak{g}}, 4)$ given in Equations 11.1 and 11.2 identify these vector spaces with $\mathbf{R}^{n-1}$ and $\mathbf{R}^{n-2}$, respectively. Via these identifications, the projections of $Q^\vee$ are the full integral lattices. In each case, the Weyl group is the full isometry group of the lattice. The images under projection of the extended simple coroots of G are a set of extended simple coroots for a root system of type BC_{n-1}, resp. BC_{n-2}. Part 5 of Theorem 1.5.1 is then clear.

11.4. Chern-Simons invariants

Let $G = Spin(4n)$ and let $\overline{G} = G/\mathcal{C}G$. Fix a $\overline{G}$-bundle ξ over the three-torus whose second Stiefel-Whitney class $w(\xi)$ evaluates over the three coordinate two-tori T_{12}, T_{13}, T_{23} to give $c_0, c_1, c_2 \in \pi_1(\overline{G}) = \mathcal{C}G$ and let Ξ be an enhanced $\overline{G}$-bundle whose underlying bundle is ξ. If $\mathbf{x} = (x, y, z)$ is a C-triple, then, by Corollary 10.2.6, there is a flat connection on Ξ whose G-holonomy around the coordinate circles is the conjugacy class of (x, y, z). Let Γ_1 be a flat connection on Ξ whose G-holonomy is the conjugacy class of a C-triple $\mathbf{x} = (x, y, z)$ in $L_{\mathcal{C}}$ with $\epsilon(\mathbf{x}) = 1$, i.e. a C-triple whose conjugacy class lies in the component X_1 of $\mathcal{T}_G(C)$ of order one. Write $\mathbf{x} = \prod_i \mathbf{x}_i \cdot \mathbf{x}_+ \cdot \mathbf{x}_-$, where $\mathbf{x}_i = (x_i, y_i, z_i)$ lies in L_i, and similarly for $\mathbf{x}_\pm$. Given a flat connection A on Ξ, $\mathrm{CS}_{\Gamma_1}(A)$ is independent mod $\mathbf{Z}$ of the choice of Γ_1, by Lemma 10.2.8 and Corollary 10.2.11. We will denote its class in $\mathbf{R}/\mathbf{Z}$ by $\mathrm{CS}(A)$.

LEMMA 11.4.1. *Let Γ be a flat connection on Ξ. If the G-holonomy of Γ is contained in X_1, then* $\mathrm{CS}(\Gamma) = 0 \bmod \mathbf{Z}$. *If the G-holonomy of Γ is contained in the component of order* 2, *then* $\mathrm{CS}(\Gamma) = 1/2 \bmod \mathbf{Z}$.

PROOF. Since CS is constant on components of $\mathcal{T}_G(C)$, the first statement is clear. To prove the second it suffices to compute $\mathrm{CS}(\Gamma)$ for one flat connection Γ whose G-holonomy is of order 2. By Lemma 11.3.4, a C-triple $\mathbf{x}'$ of order 2 is given by replacing x_1 in $\mathbf{x}$ by $\gamma_1 x_1$ where γ_1 is the non-trivial element in $\mathcal{C}L_1$. Since the inclusion of each of the A_1-factors of $L_{\mathcal{C}}$ into G induces an isomorphism on π_3, to compute the Chern-Simons invariant $\mathrm{CS}(\Gamma)$, it suffices to compute the relative Chern-Simons of the A_1-connections obtained taking the images of Γ_1 and Γ under projection the first A_1-factor of $L_{\mathcal{C}}$. Lemma 10.1.7 and Corollary 10.3.5 applied to $SU(2)$ show that this relative Chern-Simons invariant is $1/2 \bmod \mathbf{Z}$. This shows that $\mathrm{CS}(\Gamma) = 1/2 \bmod \mathbf{Z}$. □

LEMMA 11.4.2. *Let Γ be a flat connection on Ξ whose G-holonomy has order 4. Then $\mathrm{CS}(\Gamma) = \pm 1/4$ modulo $\mathbf{Z}$. If Γ' represents a point in the other component of $\mathcal{T}_G(C)$ of order 4, then $\mathrm{CS}(\Gamma') = -\mathrm{CS}(\Gamma)$.*

PROOF. Let $\mathbf{u} = (u, v, w) \in L_{I'}$ be a rank zero C-triple. Write $\mathbf{u} = \prod_{i=0}^{n-2} \mathbf{u}_i$, where $\mathbf{u}_i = (u_i, v_i, w_i)$ is a rank zero $\pi_i(C)$-triple in L_i. Let Γ be a flat connection on Ξ whose G-holonomy is the conjugacy class of $\mathbf{u}$. The C-triple $\mathbf{u}'$ obtained by replacing $\mathbf{u}_i$ by $\mathbf{u}_i' = (u_i, v_i, w_i^{-1})$ is also of rank zero. Clearly, for $i > 0$, $\delta_i(\mathbf{u}_i, \mathbf{u}_i') = 1$. By Lemma 11.1.1, $\delta_0(\mathbf{u}_0, \mathbf{u}_0') = -1$, and hence, by Lemma 11.3.7, the conjugacy classes of $\mathbf{u}'$ and $\mathbf{u}$ lie in different components of order 4 of $\mathcal{T}_G(C)$. On the other hand, given the C-triple $\mathbf{x}$ defined above, let $\mathbf{x}' = \prod_i \mathbf{x}_i' \cdot \mathbf{x}_+' \cdot \mathbf{x}_-'$, where $\mathbf{x}_i' = (x_i, y_i, z_i^{-1})$, and similarly for $\mathbf{x}_\pm'$. Clearly

$$\epsilon(\mathbf{x}) = \epsilon_+(\gamma_+)\epsilon_-(\gamma_-)\epsilon(\mathbf{x}') = \epsilon(\mathbf{x}').$$

(Here $\gamma_\pm$ are the nontrivial elements of $\mathcal{C}L_\pm$.) Thus by Lemma 11.3.4, $\mathbf{x}$ and $\mathbf{x}'$ represent points in the component X_1.

Let r be the diffeomorphism of the three-torus which is the identity on the first two factors and is inversion on the third. Note that $r(X) = X$ so that $r^*\Xi$ is an enhanced bundle over (T^3, X). Then $\mathbf{x}'$ is the G-holonomy of $r^*\Gamma_1$ on $r^*\Xi$, and hence by the above computations $r^*\Gamma_1$ represents a point in X_1. Likewise $r^*(\Gamma)$ represents a point in a different component from Γ. Since r is an orientation-reversing diffeomorphism, we have

$$\mathrm{CS}(\Gamma) \equiv \mathrm{CS}_{\Gamma_1}(\Gamma) = -\mathrm{CS}_{r^*\Gamma_1}(r^*(\Gamma)) \equiv -\mathrm{CS}(r^*(\Gamma)) \bmod \mathbf{Z}.$$

Thus, CS takes opposite values modulo $\mathbf{Z}$ on the two components of order 4 of $\mathcal{T}_G(C)$.

With $\mathbf{u}$ as above, let $\hat{\mathbf{u}}$ be the triple (u, v, w^2). Then $\hat{\mathbf{u}}$ is a c_0-triple. Likewise, let $\hat{\mathbf{x}}$ be the c_0-triple (x, y, z^2).

CLAIM 11.4.3. *The c_0-triple $\hat{\mathbf{u}}$ is in the non-trivial component of the modulo space of c_0-triples in G. The c_0-triple $\hat{\mathbf{x}}$ is in the trivial component of the modulo space of c_0-triples in G.*

PROOF. It follows from the proof of Lemma 11.1.1 that the square of w_0 in L_0 is γ_+ which is in the non-trivial component of $T_{L_0}^{w_{c_0}}$. Direct computation shows that $S^{w_{c_0}} \cap L_0$ is connected, and hence is equal to $(T_{L_0}^{w_{c_0}})^0$. Thus w^2 lies in the non-trivial component of $T^{w_{c_0}}$. Since $\pi_0(T^{w_{c_0}}) \cong \pi_0(Z(u, v))$, by Proposition 7.5.5, the c_0-triple $\hat{\mathbf{u}}$ represents a point in the non-trivial component of the moduli space of c_0-triples in G.

By Lemma 11.3.3, $z^2 = \gamma_1 \cdots \gamma_{n-1} \cdot \gamma_+ \cdot \gamma_-$. Since the $\gamma_i, 1 \leq i \leq n-1$ lie in $S^{w_{c_0}}$, it follows that z^2 is congruent modulo $S^{w_{c_0}}$ to $\gamma_+ \cdot \gamma_- = c_0$ which lies in $S^{w_{c_0}}$. Hence $z^2 \in S^{w_{c_0}}$, and so the c_0-triple $\hat{\mathbf{x}}$ lies in the trivial component of the moduli space of c_0-triples in G. □

Let t be the map on the three-torus which double covers the last coordinate. Note that t induces a map from the pair (T^3, X) to itself. Thus $t^*\Xi$ is an enhanced $\overline{G}$-bundle. The c_0-triple $\hat{\mathbf{x}}$ is the G-holonomy of $t^*\Gamma_1$ while $\hat{\mathbf{u}}$ is the G-holonomy of $t^*\Gamma$. Since the Chern-Simons invariant is given by integrating a local expression involving connections and curvature,

$$\mathrm{CS}_{t^*(\Gamma_1)}(t^*\Gamma) = 2\mathrm{CS}_{\Gamma_1}(\Gamma) \equiv 2\mathrm{CS}(\Gamma) \bmod \mathbf{Z}.$$

Since the conjugacy classes of $\hat{\mathbf{u}}$ and $\hat{\mathbf{x}}$ lie in opposite components of the moduli space of c_0-triples, by Theorem 1.8.1 applied to the case of c_0-triples,

$$\mathrm{CS}_{t^*(\Gamma_1)}(t^*\Gamma) \equiv 1/2 \bmod \mathbf{Z}.$$

It follows that

$$\mathrm{CS}(\Gamma) = \pm 1/4 \bmod \mathbf{Z}.$$

From the fact that the value of this invariant on one component of order 4 is the negative of its value on the other such component, it now follows that on one of these components the value is $1/4$ modulo $\mathbf{Z}$ and on the other it is $-1/4$ modulo $\mathbf{Z}$. □

COROLLARY 11.4.4. *For every positive integer k dividing 4, the function* CS *defines a bijection between the components of order k of $\mathcal{T}_G(C)$ and the points of order k in $\mathbf{R}/\mathbf{Z}$.*

11.5. Proof of Theorem 1.8.1 and Theorem 1.8.2 when $\langle C \rangle$ is not cyclic

Theorem 1.8.1 in the case where $\langle C \rangle$ is not cyclic is contained in Corollary 11.4.4. Lemma 11.3.1 and Corollaries 11.3.6 and 11.3.9 determine the number of components of $\mathcal{T}_G(C)$ and their dimensions. These together with Corollary 11.4.4 now give all the necessary computations in order to apply Theorem 3.8.7 to establish Witten's clockwise symmetry, Theorem 1.8.2, in the case when $\langle C \rangle$ is non-cyclic.

Bibliography

[1] A. Borel, *Sous-groupes commutatifs et torsion des groupes de Lie compactes*, Tôhoku Math. Jour. **13** (1962), 216–240.

[2] A. Borel and G. Mostow, *On semi-simple automorphisms of Lie algebras*, Annals of Math. (2), **61** (1955), 389–405.

[3] R. Bott, *An application of the Morse theory to the topology of Lie-groups*, Bull. Soc. Math. France **84** (1956), 251–281.

[4] N. Bourbaki, *Groupes et Algèbres de Lie*, Chap. 4, 5, et 6, Masson, Paris, 1981.

[5] S.-S. Chern and J. Simons, *Characteristic forms and geometric invariants*, Annals of Math. (2) **99** (1974), 48–69.

[6] J. Fuchs, U. Ray, and C. Schweigert, *Some automorphisms of generalized Kac-Moody algebras*, J. Algebra **191** (1997), 518–540.

[7] J. Fuchs, B. Schellekens, and C. Schweigert, *From Dynkin diagram symmetries to fixed point structures*, Comm. Math. Phys. **180** (1996), 39–97.

[8] R. Griess, *Elementary abelian p-subgroups of algebraic groups*, Geom. Dedicata **39** (1991), 253–305.

[9] V. Kac, *Automorphisms of finite order of semi-simple Lie algebras*, Funkt. Analis i ego Prilozh. **3** (1969), 94–96. English translation: Funct. Anal. Appl. **3** (1969), 252–254.

[10] V. Kac, *Infinite Dimensional Lie Algebras*, Third Edition, Cambridge University Press, Cambridge, 1990.

[11] V. Kac and A. Smilga, *Vacuum structure in supersymmetric Yang-Mills theories with any gauge group*, hep-th/9902029 v. 3.

[12] A. Keurentjes, *Non-trivial flat connections on the 3-torus I: G_2 and the orthogonal groups*, hep-th/9901154, J. High Energy Phys. **9905** (1999), 001 (electronic).

[13] A. Keurentjes, *Non-trivial flat connections on the 3-torus II: The exceptional groups F_4 and $E_{6,7,8}$*, hep-th/9902186, J. High Energy Phys. **9905** (1999), 014 (electronic).

[14] E. Looijenga, *Root systems and elliptic curves*, Invent. Math. **38** (1976), 17–32.

[15] M. Mimura and H. Toda, *Topology of Lie Groups I,II*, Translations of Mathematical Monographs **91**, American Mathematical Society, Providence, 1991.

[16] C. Schweigert, *On moduli spaces of flat connections with non-simply connected structure group*, Nuclear Phys. B **492** (1997), 743–755.

[17] J. de Siebenthal, *Sur les groupes de Lie compacts non connexes*, Comm. Math. Helv. **31** (1956/57), 41–89.

[18] R. Steinberg, *Endomorphisms of linear algebraic groups*, Memoirs of the AMS **80**, American Mathematical Society, Providence, 1968.

[19] R. Steinberg, *Torsion in reductive groups*, Advances in Math. **15** (1975), 63–92.

[20] E. Witten, *Toroidal compactification without vector structure*, J. High Energy Phys. **9802** (1998), 006 (electronic).

Diagrams and tables

Extended coroot diagrams and extended coroot integers

In the diagrams, ∘ represents the extended coroot.

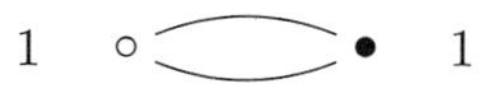

$\widetilde{A}_1^\vee = \widetilde{A}_1$

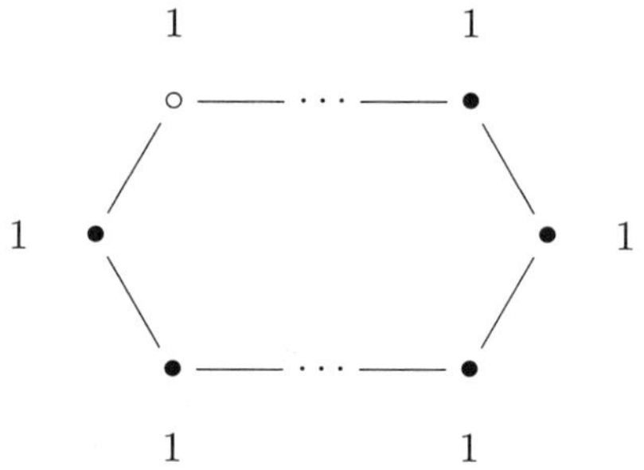

$\widetilde{A}_n^\vee = \widetilde{A}_n,\ n \geq 2$

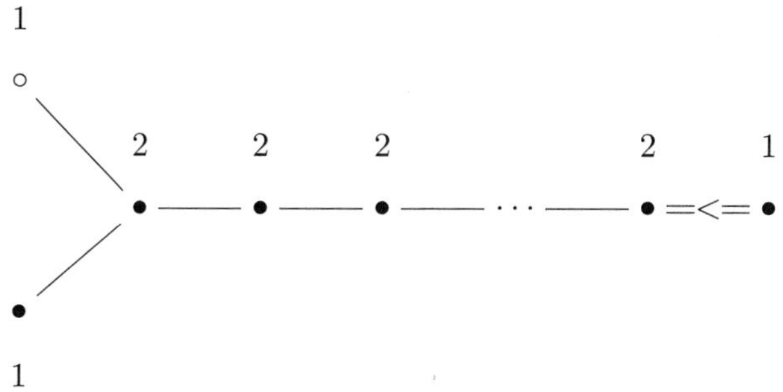

$\widetilde{B}_n^\vee,\ n \geq 3$

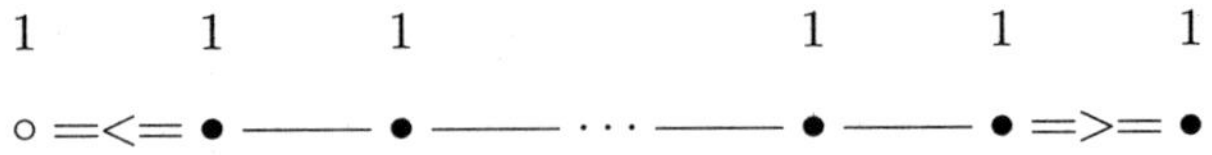

$\widetilde{C}_n^\vee$, $n \geq 2$

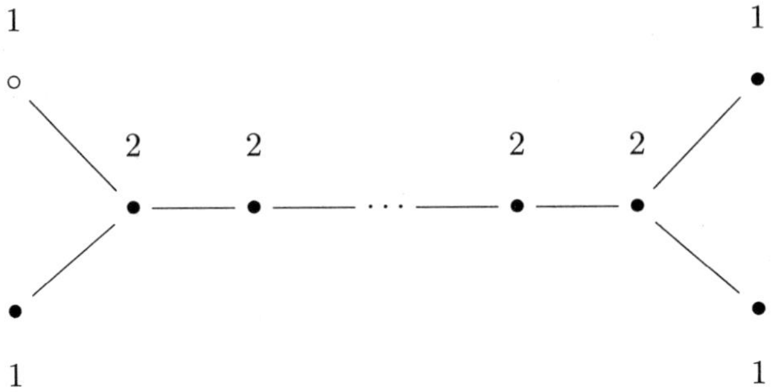

$\widetilde{D}_n^\vee = \widetilde{D}_n$, $n \geq 4$

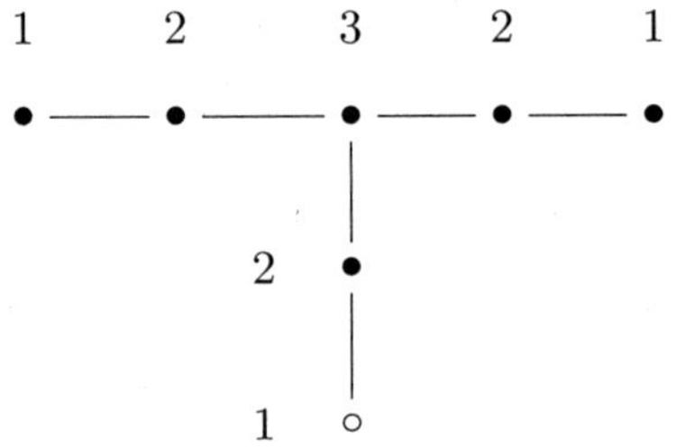

$\widetilde{E}_6^\vee = \widetilde{E}_6$

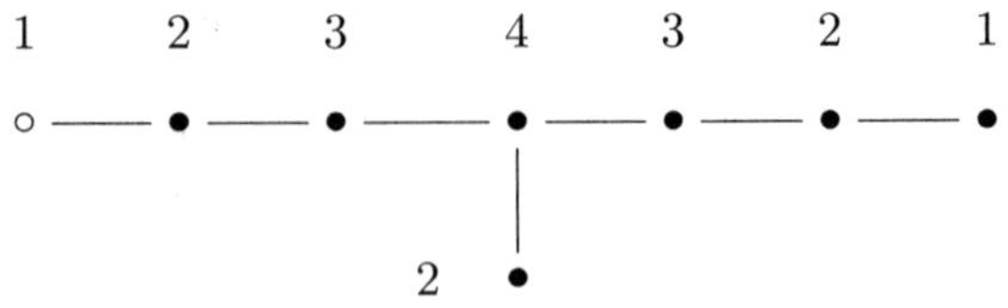

$\widetilde{E}_7^\vee = \widetilde{E}_7$

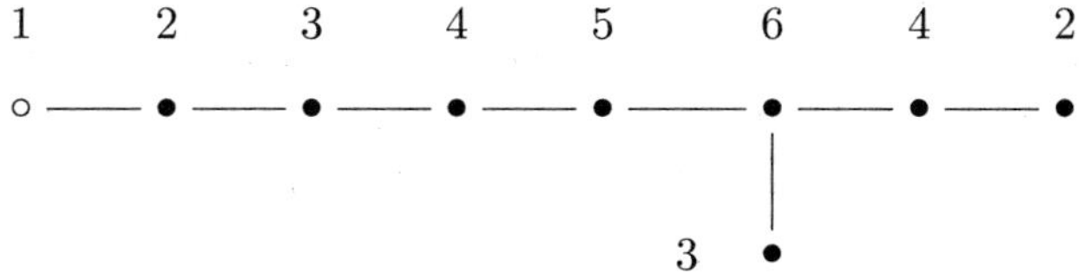

$\widetilde{E}_8^\vee = \widetilde{E}_8$

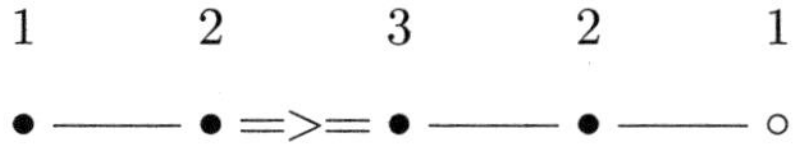

$\widetilde{F}_4^\vee$

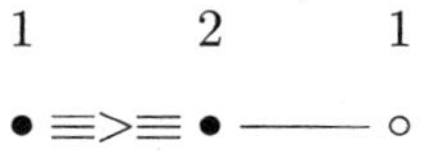

$\widetilde{G}_2^\vee$

$\widetilde{BC}_1^\vee$

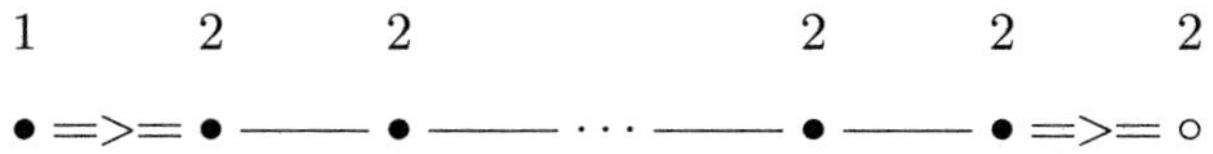

$\widetilde{BC}_n^\vee$, $n \geq 2$

Quotient extended coroot diagrams and quotient coroot integers

In the diagrams below, c denotes an element of the center and $o(c)$ is its order. Except in the case of $\widetilde{A}_n$, c will always denote a generator of the center. c_{SO} is the non-trivial element in $\pi_1(SO(2n)) \subset \mathcal{C}Spin(2n)$ and c_{exotic} is any other non-trivial element in $\mathcal{C}Spin(4n)$.

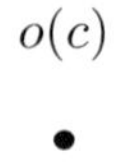

$\widetilde{A}_n^\vee/c$ when $o(c) = n+1$

$\widetilde{A}_n^\vee/c$ when $o(c) = (n+1)/2$

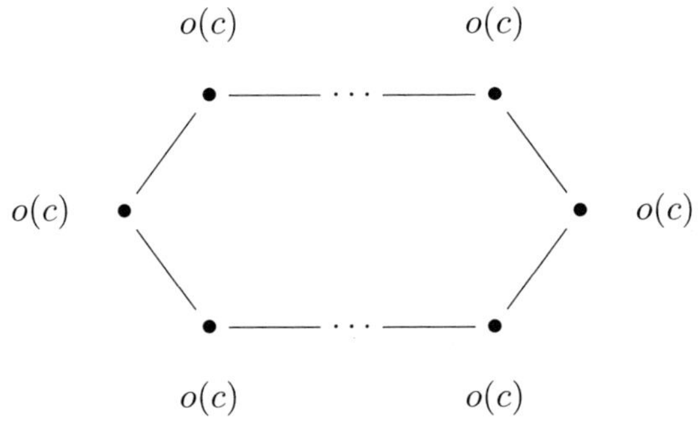

$\widetilde{A}_n^\vee/c$ when $o(c) < (n+1)/2$

$\widetilde{B}_n^\vee/c$

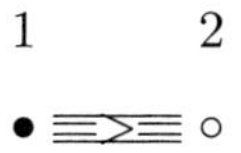

$$C_2^\vee/c$$

$$\overset{1}{\bullet}\!=\!>\!=\!\overset{2}{\bullet}\!-\!\!-\!\overset{2}{\bullet}\!-\!\!-\cdots-\!\!-\overset{2}{\bullet}\!-\!\!-\overset{2}{\bullet}\!=\!>\!=\!\overset{2}{\bullet}$$
$$\widetilde{C}_{2n}^\vee/c,\ n\geq 2$$

$$\overset{2}{\bullet}\!\!\overset{\frown}{\underset{\smile}{}}\!\!\overset{2}{\bullet}$$
$$\widetilde{C}_3^\vee/c$$

$$\overset{2}{\bullet}\!=\!<\!=\!\overset{2}{\bullet}\!-\!\!-\!\overset{2}{\bullet}\!-\!\!-\cdots-\!\!-\overset{2}{\bullet}\!-\!\!-\overset{2}{\bullet}\!=\!>\!=\!\overset{2}{\bullet}$$
$$\widetilde{C}_{2n+1}^\vee/c,\ n\geq 2$$

$$\overset{2}{\bullet}\!=\!<\!=\!\overset{2}{\bullet}\!-\!\!-\!\overset{2}{\bullet}\!-\!\!-\cdots-\!\!-\overset{2}{\bullet}\!-\!\!-\overset{2}{\bullet}\!=\!>\!=\!\overset{2}{\bullet}$$
$$\widetilde{D}_n^\vee/c_{\mathrm{SO}}$$

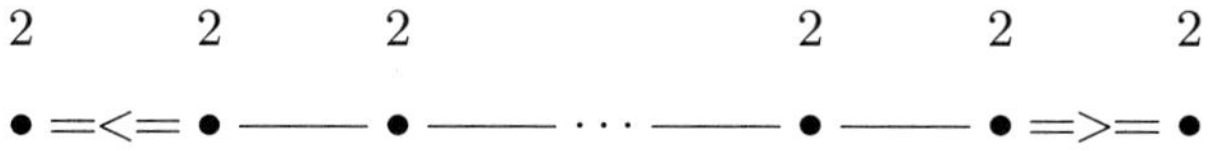

$$\overset{2}{\bullet}=\!<\!=\overset{2}{\bullet}=\!>\!=\overset{2}{\bullet}$$
$$D_4^\vee/c_{\text{exotic}}$$

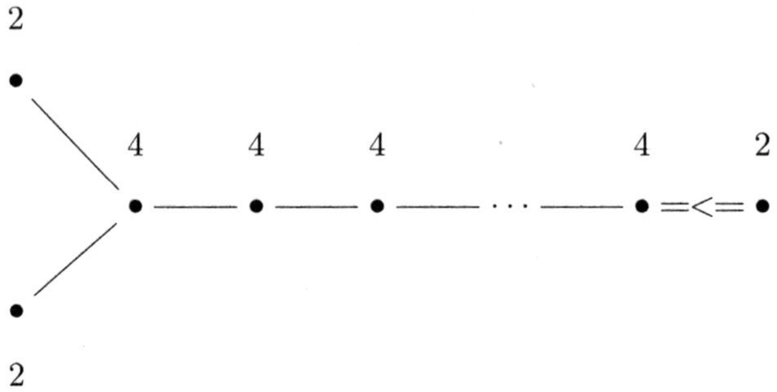

$$\widetilde{D}_{2n}^\vee/c_{\text{exotic}},\ n\geq 3$$

$$\overset{4}{\bullet}=\!<\!=\overset{4}{\bullet}\!-\!\!-\!\overset{4}{\bullet}\!-\!\!-\cdots\!-\!\!-\overset{4}{\bullet}\!-\!\!-\overset{4}{\bullet}=\!>\!=\overset{4}{\bullet}$$
$$\widetilde{D}_{2n+1}^\vee/c$$

$$\overset{2}{\bullet}\equiv\!>\!\equiv\overset{4}{\bullet}$$
$$D_4/\mathcal{C}D_4$$

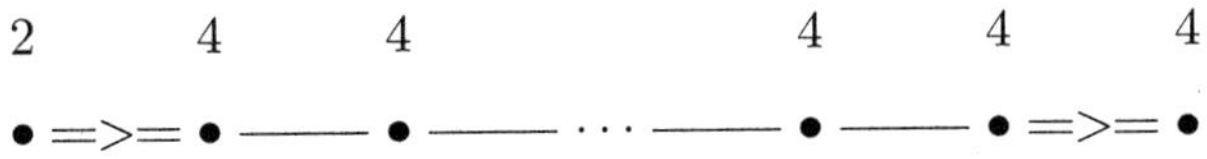

$\widetilde{D}^{\vee}_{2n}/\mathcal{C}D_{2n},\ n \geq 3$

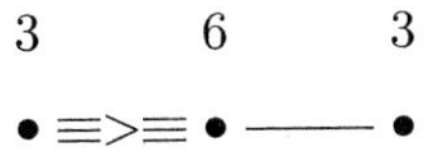

$\widetilde{E}^{\vee}_{6}/c$

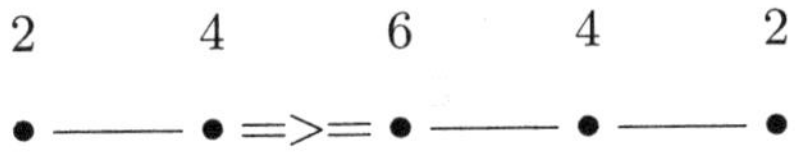

$\widetilde{E}^{\vee}_{7}/c$

Root systems on $\mathfrak{t}^{w_{\mathcal{C}}}$

G	$\mathcal{C}$	$L_{\mathcal{C}}$	$\Phi^{w_{\mathcal{C}}}$	Φ^{res}	Φ^{proj}	$\Phi(w_{\mathcal{C}})$	$g_{\overline{a}}$
B_n	$\mathcal{C}B_n$	A_1 (short root)	C_{n-1}	B_{n-1}	BC_{n-1}	BC_{n-1}	$1, 2, \ldots, 2$
C_{2n+1}	$\mathcal{C}C_{2n+1}$	$\prod_{i=1}^{n+1} A_1$	C_n	BC_n	BC_n	C_n	$2, 2, \ldots, 2$
C_{2n}	$\mathcal{C}C_{2n}$	$\prod_{i=1}^{n} A_1$	C_n	C_n	BC_n	BC_n	$1, 2, \ldots, 2$
D_n	$\pi_1(SO(2n))$	$A_1 \times A_1$	C_{n-2}	B_{n-2}	C_{n-2}	C_{n-2}	$2, 2, \ldots, 2$
D_{2n}	$\langle c_{\mathrm{exotic}} \rangle$	$\prod_{i=1}^{n} A_1$	B_n	C_n	B_n	B_n	$2, 4, \ldots, 4, 2, 2$
D_{2n+1}	$\mathcal{C}D_{2n+1}$	$\prod_{i=1}^{n-1} A_1 \times A_3$	C_{n-1}	BC_{n-1}	BC_{n-1}	C_{n-1}	$4, 4, \ldots, 4$
D_{2n}	$\mathcal{C}D_{2n}$	$\prod_{i=1}^{n+1} A_1$	C_{n-1}	BC_{n-1}	BC_{n-1}	BC_{n-1}	$2, 4, \ldots, 4$
E_6	$\mathcal{C}E_6$	$A_2 \times A_2$	G_2	G_2	G_2	G_2	$3, 6, 3$
E_7	$\mathcal{C}E_7$	$A_1 \times A_1 \times A_1$	F_4	F_4	F_4	F_4	$2, 4, 6, 4, 2$

Root systems on $\mathfrak{t}(k)$ for $k > 1$

G	k	L	$\Phi(\mathfrak{t}(k))$	g_a divisible by k
B_n	2	B_3	C_{n-3}	$2, 2, \ldots, 2$
D_n	2	D_4	C_{n-4}	$2, 2, \ldots, 2$
E_6	2	D_4	A_2	$2, 2, 2$
E_6	3	E_6	trivial	3
E_7	2	D_4	B_3	$2, 4, 2, 2$
E_7	3	E_6	A_1	$3, 3$
E_7	4	E_7	trivial	4
E_8	2	D_4	F_4	$2, 4, 6, 4, 2$
E_8	3	E_6	G_2	$3, 6, 3$
E_8	4	E_7	A_1	$4, 4$
E_8	5	E_8	trivial	5
E_8	6	E_8	trivial	6
F_4	2	B_3	A_1	$2, 2$
F_4	3	F_4	trivial	3
G_2	2	G_2	trivial	2

Root systems on $\mathfrak{t}^{w_C}(\overline{\mathfrak{g}}, k)$ for $\langle C \rangle \neq 1$ and $k \nmid n_0$

G	$\langle C \rangle$	k	L	$\Phi(\mathfrak{t}^{w_C}(\overline{\mathfrak{g}}, k))$	$g_{\overline{a}}$ divisible by k
B_n	$\mathcal{C}B_n$	2	C_2	C_{n-2}	$2, 2, \ldots, 2$
C_{2n}	$\mathcal{C}C_{2n}$	2	$\prod_{i=1}^{n-1} A_1 \times C_2$	C_{n-1}	$2, 2, \ldots, 2$
$D_{2n}, n \geq 3$	$\langle c_{\text{exotic}} \rangle$	4	$\prod_{i=1}^{n-3} A_1 \times D_6$	C_{n-3}	$4, 4, \ldots, 4$
D_{2n}	$\mathcal{C}D_{2n}$	4	$\prod_{i=1}^{n-2} A_1 \times D_4$	C_{n-2}	$4, 4 \ldots, 4$
E_6	$\mathcal{C}E_6$	2	E_6	trivial	6
E_6	$\mathcal{C}E_6$	6	E_6	trivial	6
E_7	$\mathcal{C}E_7$	4	D_6	A_1	$4, 4$
E_7	$\mathcal{C}E_7$	3	E_7	trivial	6
E_7	$\mathcal{C}E_7$	6	E_7	trivial	6

Editorial Information

To be published in the *Memoirs*, a paper must be correct, new, nontrivial, and significant. Further, it must be well written and of interest to a substantial number of mathematicians. Piecemeal results, such as an inconclusive step toward an unproved major theorem or a minor variation on a known result, are in general not acceptable for publication. Papers appearing in *Memoirs* are generally longer than those appearing in *Transactions*, which shares the same editorial committee.

As of January 31, 2002, the backlog for this journal was approximately 5 volumes. This estimate is the result of dividing the number of manuscripts for this journal in the Providence office that have not yet gone to the printer on the above date by the average number of monographs per volume over the previous twelve months, reduced by the number of volumes published in four months (the time necessary for preparing a volume for the printer). (There are 6 volumes per year, each containing at least 4 numbers.)

A Consent to Publish and Copyright Agreement is required before a paper will be published in the *Memoirs*. After a paper is accepted for publication, the Providence office will send a Consent to Publish and Copyright Agreement to all authors of the paper. By submitting a paper to the *Memoirs*, authors certify that the results have not been submitted to nor are they under consideration for publication by another journal, conference proceedings, or similar publication.

Information for Authors

Memoirs are printed from camera copy fully prepared by the author. This means that the finished book will look exactly like the copy submitted.

The paper must contain a *descriptive title* and an *abstract* that summarizes the article in language suitable for workers in the general field (algebra, analysis, etc.). The *descriptive title* should be short, but informative; useless or vague phrases such as "some remarks about" or "concerning" should be avoided. The *abstract* should be at least one complete sentence, and at most 300 words. Included with the footnotes to the paper should be the 2000 *Mathematics Subject Classification* representing the primary and secondary subjects of the article. The classifications are accessible from `www.ams.org/msc/`. The list of classifications is also available in print starting with the 1999 annual index of *Mathematical Reviews*. The Mathematics Subject Classification footnote may be followed by a list of *key words and phrases* describing the subject matter of the article and taken from it. Journal abbreviations used in bibliographies are listed in the latest *Mathematical Reviews* annual index. The series abbreviations are also accessible from `www.ams.org/publications/`. To help in preparing and verifying references, the AMS offers MR Lookup, a Reference Tool for Linking, at `www.ams.org/mrlookup/`. When the manuscript is submitted, authors should supply the editor with electronic addresses if available. These will be printed after the postal address at the end of the article.

Electronically prepared manuscripts. The AMS encourages electronically prepared manuscripts, with a strong preference for $\mathcal{AMS}$-LaTeX. To this end, the Society has prepared $\mathcal{AMS}$-LaTeX author packages for each AMS publication. Author packages include instructions for preparing electronic manuscripts, the *AMS Author Handbook*, samples, and a style file that generates the particular design specifications of that publication series. Though $\mathcal{AMS}$-LaTeX is the highly preferred format of TeX, author packages are also available in $\mathcal{AMS}$-TeX.

Authors may retrieve an author package from e-MATH starting from `www.ams.org/tex/` or via FTP to `ftp.ams.org` (login as `anonymous`, enter username as password, and type `cd pub/author-info`). The *AMS Author Handbook* and the *Instruction Manual* are available in PDF format following the author packages link from `www.ams.org/tex/`. The author package can be obtained free of charge by sending email to `pub@ams.org` (Internet) or from the Publication Division, American Mathematical Society, P.O. Box 6248, Providence, RI 02940-6248. When requesting an author package, please specify AMS-LaTeX or AMS-TeX, Macintosh or IBM (3.5) format, and the publication in which your paper will appear. Please be sure to include your complete mailing address.

Sending electronic files. After acceptance, the source file(s) should be sent to the Providence office (this includes any TeX source file, any graphics files, and the DVI or PostScript file).

Before sending the source file, be sure you have proofread your paper carefully. The files you send must be the EXACT files used to generate the proof copy that was accepted for publication. For all publications, authors are required to send a printed copy of their paper, which exactly matches the copy approved for publication, along with any graphics that will appear in the paper.

TeX files may be submitted by email, FTP, or on diskette. The DVI file(s) and PostScript files should be submitted only by FTP or on diskette unless they are encoded properly to submit through email. (DVI files are binary and PostScript files tend to be very large.)

Electronically prepared manuscripts can be sent via email to `pub-submit@ams.org` (Internet). The subject line of the message should include the publication code to identify it as a Memoir. TeX source files, DVI files, and PostScript files can be transferred over the Internet by FTP to the Internet node `e-math.ams.org` (130.44.1.100).

Electronic graphics. Comprehensive instructions on preparing graphics are available at `www.ams.org/jourhtml/graphics.html`. A few of the major requirements are given here.

Submit files for graphics as EPS (Encapsulated PostScript) files. This includes graphics originated via a graphics application as well as scanned photographs or other computer-generated images. If this is not possible, TIFF files are acceptable as long as they can be opened in Adobe Photoshop or Illustrator. No matter what method was used to produce the graphic, it is necessary to provide a paper copy to the AMS.

Authors using graphics packages for the creation of electronic art should also avoid the use of any lines thinner than 0.5 points in width. Many graphics packages allow the user to specify a "hairline" for a very thin line. Hairlines often look acceptable when proofed on a typical laser printer. However, when produced on a high-resolution laser imagesetter, hairlines become nearly invisible and will be lost entirely in the final printing process.

Screens should be set to values between 15% and 85%. Screens which fall outside of this range are too light or too dark to print correctly. Variations of screens within a graphic should be no less than 10%.

Inquiries. Any inquiries concerning a paper that has been accepted for publication should be sent directly to the Electronic Prepress Department, American Mathematical Society, P. O. Box 6248, Providence, RI 02940-6248.

Editors

This journal is designed particularly for long research papers, normally at least 80 pages in length, and groups of cognate papers in pure and applied mathematics. Papers intended for publication in the *Memoirs* should be addressed to one of the following editors. In principle the Memoirs welcomes electronic submissions, and some of the editors, those whose names appear below with an asterisk (*), have indicated that they prefer them. However, editors reserve the right to request hard copies after papers have been submitted electronically. Authors are advised to make preliminary email inquiries to editors about whether they are likely to be able to handle submissions in a particular electronic form.

Algebra to KAREN E. SMITH, Department of Mathematics, University of Michigan, 525 University, Suite 2832, Ann Arbor, MI 48109-1109; email: kesmith@lsa.umich.edu

Algebraic geometry and commutative algebra to LAWRENCE EIN, Department of Mathematics, University of Illinois, 851 S. Morgan (M/C 249), Chicago, IL 60607-7045; email: ein@uic.edu

Algebraic topology and cohomology of groups to STEWART PRIDDY, Department of Mathematics, Northwestern University, 2033 Sheridan Road, Evanston, IL 60208-2730; email: priddy@math.nwu.edu

Combinatorics and Lie theory to SERGEY FOMIN, Department of Mathematics, University of Michigan, Ann Arbor, Michigan 48109-1109; email: fomin@math.lsa.umich.edu

Complex analysis and complex geometry to DUONG H. PHONG, Department of Mathematics, Columbia University, 2990 Broadway, New York, NY 10027-0029; email: phong@math.columbia.edu

*__Differential geometry and global analysis__ to LISA C. JEFFREY, Department of Mathematics, University of Toronto, 100 St. George St., Toronto, ON Canada M5S 3G3; email: jeffrey@math.toronto.edu

Dynamical systems and ergodic theory to ROBERT F. WILLIAMS, Department of Mathematics, University of Texas, Austin, Texas 78712-1082; email: bob@math.utexas.edu

Functional analysis and operator algebras to DAN VOICULESCU, Department of Mathematics, University of California, Berkeley, 970 Evans Hall, Floor 9, Berkeley, CA 94720-0001; email: dvv@math.berkeley.edu

Geometric topology, knot theory and hyperbolic geometry to ABIGAIL A. THOMPSON, Department of Mathematics, University of California, Davis, Davis, CA 95616-5224; email: thompson@math.ucdavis.edu

Harmonic analysis, representation theory, and Lie theory to ROBERT J. STANTON, Department of Mathematics, The Ohio State University, 231 West 18th Avenue, Columbus, OH 43210-1174; email: stanton@math.ohio-state.edu

*__Logic__ to THEODORE SLAMAN, Department of Mathematics, University of California, Berkeley, CA 94720-3840; email: slaman@math.berkeley.edu

Number theory to HAROLD G. DIAMOND, Department of Mathematics, University of Illinois, 1409 W. Green St., Urbana, IL 61801-2917; email: diamond@math.uiuc.edu

*__Ordinary differential equations, partial differential equations, and applied mathematics__ to PETER W. BATES, Department of Mathematics, Michigan State University, East Lansing, MI 48824-1027; email: bates@math.msu.edu

*__Probability and statistics__ to KRZYSZTOF BURDZY, Department of Mathematics, University of Washington, Box 354350, Seattle, Washington 98195-4350; email: burdzy@math.washington.edu

*__Real and harmonic analysis and geometric partial differential equations__ to WILLIAM BECKNER, Department of Mathematics, University of Texas, Austin, TX 78712-1082; email: beckner@math.utexas.edu

All other communications to the editors should be addressed to the Managing Editor, WILLIAM BECKNER, Department of Mathematics, University of Texas, Austin, TX 78712-1082; email: beckner@math.utexas.edu.

Selected Titles in This Series

(Continued from the front of this publication)

718 **Bernhard Lani-Wayda,** Wandering solutions of delay equations with sine-like feedback, 2001

717 **Ron Brown,** Frobenius groups and classical maximal orders, 2001

716 **John H. Palmieri,** Stable homotopy over the Steenrod algebra, 2001

715 **W. N. Everitt and L. Markus,** Multi-interval linear ordinary boundary value problems and complex symplectic algebra, 2001

714 **Earl Berkson, Jean Bourgain, and Aleksander Pełczynski,** Canonical Sobolev projections of weak type $(1,1)$, 2001

713 **Dorina Mitrea, Marius Mitrea, and Michael Taylor,** Layer potentials, the Hodge Laplacian, and global boundary problems in nonsmooth Riemannian manifolds, 2001

712 **Raúl E. Curto and Woo Young Lee,** Joint hyponormality of Toeplitz pairs, 2001

711 **V. G. Kac, C. Martinez, and E. Zelmanov,** Graded simple Jordan superalgebras of growth one, 2001

710 **Brian Marcus and Selim Tuncel,** Resolving Markov chains onto Bernoulli shifts via positive polynomials, 2001

709 **B. V. Rajarama Bhat,** Cocylces of CCR flows, 2001

708 **William M. Kantor and Ákos Seress,** Black box classical groups, 2001

707 **Henning Krause,** The spectrum of a module category, 2001

706 **Jonathan Brundan, Richard Dipper, and Alexander Kleshchev,** Quantum Linear groups and representations of $GL_n(\mathbb{F}_q)$, 2001

705 **I. Moerdijk and J. J. C. Vermeulen,** Proper maps of toposes, 2000

704 **Jeff Hooper, Victor Snaith, and Min van Tran,** The second Chinburg conjecture for quaternion fields, 2000

703 **Erik Guentner, Nigel Higson, and Jody Trout,** Equivariant E-theory for C^*-algebras, 2000

702 **Ilijas Farah,** Analytic guotients: Theory of liftings for quotients over analytic ideals on the integers, 2000

701 **Paul Selick and Jie Wu,** On natural coalgebra decompositions of tensor algebras and loop suspensions, 2000

700 **Vicente Cortés,** A new construction of homogeneous quaternionic manifolds and related geometric structures, 2000

699 **Alexander Fel′shtyn,** Dynamical zeta functions, Nielsen theory and Reidemeister torsion, 2000

698 **Andrew R. Kustin,** Complexes associated to two vectors and a rectangular matrix, 2000

697 **Deguang Han and David R. Larson,** Frames, bases and group representations, 2000

696 **Donald J. Estep, Mats G. Larson, and Roy D. Williams,** Estimating the error of numerical solutions of systems of reaction-diffusion equations, 2000

695 **Vitaly Bergelson and Randall McCutcheon,** An ergodic IP polynomial Szemerédi theorem, 2000

694 **Alberto Bressan, Graziano Crasta, and Benedetto Piccoli,** Well-posedness of the Cauchy problem for $n \times n$ systems of conservation laws, 2000

693 **Doug Pickrell,** Invariant measures for unitary groups associated to Kac-Moody Lie algebras, 2000

692 **Mara D. Neusel,** Inverse invariant theory and Steenrod operations, 2000

691 **Bruce Hughes and Stratos Prassidis,** Control and relaxation over the circle, 2000